CONTENTS

THE CALENDAR
HISTORY, LORE, AND LEGEND

Jacqueline de Bourgoing

DISCOVERIES®
HARRY N. ABRAMS, INC., PUBLISHERS

"Why have calendars at all? In order to predict the regular patterns of nature. In an agricultural society you need a solar calendar to know when best to sow your crops. In a society that lives by fishing you need a lunar calendar to know the tides. Yet it's impossible to establish a simple arithmetical relationship between the two that would bring them into harmony."

Stephen Jay Gould,
from Umberto Eco, Stephen Jay Gould, et al.,
Conversations about the End of Time, 1998,
trans. Ian Maclean and Roger Pearson

CHAPTER 1
AN ANCIENT CHALLENGE

To create a calendar requires a thorough knowledge of the motions of the stars and many cumulative, learned adjustments. Left: a 1st-century BC Celtic calendar discovered at Coligny, France. Right: a 16th-century engraving of a northern European runic calendar. Both attempted to adapt a lunar system to the march of the seasons.

A tool for living in society

The Maya, Aztec, Gauls, Greeks, Romans, Chinese, Jews, Copts, Muslims—all have calendars. The urge to organize time is shared by all societies. In fact, to live together, people must create a uniform framework for referring to time that allows the coordination of human activities.

The calendar, a system of dividing time into days, months, and years, based on astronomical cycles, responds to this vital need. The calendar is a bridge between cosmic time and time as lived by each individual (to borrow the terms used by the philosopher Paul Ricoeur). It creates a social time, different from cosmic or personal time: a time that is intelligible to everyone.

It fulfills two basic functions: to give time a rhythm and means of being measured. The rhythm is created by means of a structure that differentiates work days from holidays and festivals and establishes traditions, thus forming a symbolic link among the members of a community. Every society has its own calendar, which expresses its identity. The measurement of time is a somewhat more objective task; it is done by identifying, as accurately as possible, periods such as the length of a year, a month, and smaller divisions. In other words, measuring time means imposing a certain order based on selected temporal recurrences, and then maintaining it.

The impossibility of harmonizing astronomical cycles

All calendars are based on observation of the motion of heavenly bodies, and use as their base unit a natural cycle whose regular rhythms are clear to everyone. Three astronomical cycles serve as reference points: the earth's rotation on its axis, which defines the length of a day; the passage of the moon around the earth, which defines the lunar month; and the earth's orbit around the sun, which defines the year.

Yet these three cycles pose considerable problems for the establishment of calendars: there is no simple mathematical relation between them; none of them is

"The calendar expresses the rhythm of collective life while it serves the function of ensuring its regularity."

Emile Durkheim,
The Elementary Forms of Religious Life, 1912

Arranged like a comic strip, the Aztec Codex Borbonicus was drawn in the 16th century, shortly before the arrival of the Spanish Conquistadores. It is something of a divinatory and ritual primer explaining the structure of the Aztec calendar. The page reproduced here shows the "link of years," that is, the way the Aztec, following the Maya, combined a divinatory calendar of 260 days and a civil calendar of 365 days in a cycle of 52 solar years, distributed in four groups of 13 years. Each year is indicated by the glyph symbolizing it, its number in the order of the series, and one of four images that follow one another in the same sequence: the vine, signifying the rising sun and rebirth; the knife, symbolizing the north, sacrifices, and shadows; the house, representing the west, where the sun withdraws at night; and the rabbit, designating the moon. In the center the gods Cipactonal and his wife Oxomoco, seated in a cave, discuss the making of the calendar, an essential element in the power of the gods in pre-Columbian Meso-American civilizations.

divisible by the others; and moreover their durations can vary.

Consider the year, for instance. It can be measured in various ways to obtain different durations. The stellar year corresponds to the time in which the sun returns to a position facing a given star. It lasts about 365 days, 6 hours, 9 minutes, and 9.54 seconds. The tropical year

ریب بید خسوف قمر وچون زمین حایل شود میان جرم قمر وجرم آفتاب

ذنب پدید آید و قمر در نقطهٔ راس با ذنب باشد و جرم آفتاب بیش از کرهٔ زمین است

از ظل زمین مخروطی پدید آید قاعدهٔ او سطح زمین باشد از بهر آنکه خطوط شعاعی کآفتاب

سطح زمین متساوی نباشد چون جرم زمین رسد از اصحاب وکبد ز دو بیک کر متصل شود

نقطه از سایهٔ زمین شکل مخروطی پدید آید خاک که شرح داده شد اکر قمر را عرض بود از فلک

represents the earth's revolution around the sun, from one vernal (spring) equinox to the next. At present it lasts an average of 365 days, 5 hours, 48 minutes, 45.96 seconds, but because the earth's motion is gradually slowing, the tropical year regularly loses five seconds every thousand years—a fact that makes it impossible to establish a perfect calendar in the long run.

The lunar cycle, the interval between two new moons—that is, between two alignments of the moon and the sun—is no less complicated. It lasts an *average* of 29 days, 12 hours, 44 minutes, 3 seconds, but its length is very flexible. It can be as short as 29 days, 6 hours, or as long as 29 days, 20 hours.

The duration of a day, the interval between two sunrises or sunsets, or two successive noontimes, is not regular either. It varies between 23 hours, 59 minutes, 39 seconds and 24 hours, 0 minutes, 30 seconds. The 24-hour day is an approximate average.

Neither a lunar month nor a year is made up of whole days. However, for a calendar to be useable, it must necessarily reduce time to simple units, such as days of exactly 24 hours, which do not exist in the astronomical cycles. Every society that has tried to make a calendar has run up against this difficulty. Whatever reference heavenly body is chosen, whether the moon, the sun, or another

The moon's revolution around the earth serves to measure time. A lunar month has four phases: the new moon (conjunction), when the moon is between the sun and earth; the first quarter; the full moon (opposition); and the final quarter. This spectacular series—appearance, expansion, contraction, and disappearance—played a major role in working out cyclic concepts. Left: a 16th-century Persian manuscript illustrates the phases of the moon.

Many early peoples calculated time in lunar months. Below: a 19th-century Siberian lunar calendar made of mammoth bone.

star, it is necessary to use average values and to simplify. Calendars are always homemade improvisations, compromises with complex astronomical cycles.

The first calendars were lunar

Is it better to divide time by reference to the moon or the sun? Is it possible to reconcile the lunar and solar cycles in a single calendar?

Measuring time on the basis of lunar months is a very old idea. The lunar cycle is easy to observe. Its phases can be related easily to the moon's changing shape. It is possible to establish rough equivalences with the solar year: 12 lunar months (354.36 days) correspond roughly to the sequence of seasons. In ancient Mesopotamia, Egypt, Greece, Rome, and China, people first made lunar calendars, which usually had 12 months alternating between 30 and 29 days. This alternation was due to the fact that the lunar month has 29.53 days. In Babylonia, when the disparity with the slightly longer solar year of approximately 365 days became awkward, the sovereign ordered the addition of an extra month.

Today, the best-known lunar calendar is the Muslim calendar. Adopted officially in AD 634, it consists of 12 months made up of 29 and 30 days. The last month has a variable duration, 29 or 30 days, in order to match as closely as possible the motions of the moon.

"It is he who gave its splendor to the sun and its light to the moon, whose periods he has arranged so as to allow us to count the seasons and years."

The Quran,
Sura Yunus 10:5

In Muslim time-reckoning, lunar years are integrated into cycles of 30 years, composed of 19 years of 354 days (common years) and 11 years of 355 days (abundant years). Since the Muslim year has 11 or 12 fewer days than the solar year, the months fall behind the seasons every year. The coincidence between the Muslim and Gregorian calendars recurs every 34 years. Left: a Muslim perpetual calendar.

The Prophet Muhammad instituted two calendar feasts: Aid al-Fitr, marking the end of the fast at the close of the month of Ramadan, and Aid al-Addha, the day of sacrifice. This second sacred period, which lasts three days, is also known as Beîram. In it Muslims commemorate Abraham's sacrifice by slaying an animal, usually a sheep but sometimes a cow or camel.

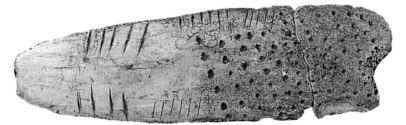

Lunar calendars are appropriate for nomadic populations and seafaring societies, but can cause problems for farm cultures, because they cannot be aligned well with the seasons. Some societies tried

Above: the hatch marks cut into this prehistoric reindeer antler may represent the phases of the moon, in a rough attempt at a calendar. It was found at Les Eyzies, in the Dordogne region of France, and dates to 30,000–25,000 BC.

In Mesopotamia the moon god Sin, portrayed in an 8th-century BC carving, was the father of the sun god, Shamash.

The Egyptians and Maya were not the only ancient peoples to base their calendar on the sun. The alignment of the megaliths of Stonehenge, in southern England, testifies to a precise knowledge of the year and its seasons on the part of the Neolithic culture there. Enormous stone blocks are arranged in concentric circles so that at the summer solstice the sun rises on the axis of the central alley, and shines directly on a flat stone that may have been an altar. Erected c. 2000 BC by a people of whom we know nothing, Stonehenge was probably both a sanctuary and an observatory used to measure time.

"The German term for 'solstice' [*Sonnenwende*] means, literally, the point where the sun changes direction; having reached the farthest point in its summer circuit, it turns back...Like many other peoples, the builders of Stonehenge tried to determine the moment when the sun reverses its motion in relation to them. They did so because they saw in this change a signal given to their group to engage in a certain kind of activity."
Norbert Elias,
Time: An Essay, 1987,
trans. by
Edmund Jephcott

to adjust their lunar calendar to the solar year by systematically adding intermediate months, called intercalar months. Thus their calendar became lunisolar.

Meton's cycle adapted by the Jews

The Greek astronomer Meton (5th century BC) is credited with the development of an effective system to adjust lunar calendars to the solar year. Meton supposedly observed that 19 solar years corresponded almost exactly to 235 lunar months. Since 19 lunar years consist of 228 months, it was thus necessary to add 7 intermediate months into 19 lunar years to put the lunar and solar calendars in phase. Impressed by this discovery, the Athenians had Meton's cycle engraved in gold letters on the temple of Athena on the occasion of the Olympic Games of 432 BC. Yet they failed to use it rigorously. The intermediate months were not always placed correctly.

In the 4th century AD, the Jews made admirable use of this cycle and established an extremely sophisticated lunisolar calendar. For religious reasons, they found it necessary to adjust their traditional lunar calendar to the rhythm of the seasons. The Passover holiday was supposed to take place at the beginning of spring because it commemorated the Exodus from Egypt, which, according to the Bible, had coincided with the spring festival of Azymes. When the months grew too far out of alignment with the seasons, the barley necessary for the Passover rite would not be ripe by the month of Nisan (the month of Passover). Then the Sanhedrin, the college of priests, worked out an empirical adjustment by doubling the last month of the year. In AD 359, Rabbi Hillel II decided to reform the calendar so that the Jews living in the Diaspora could all celebrate the holidays at the same time. At that point the Hebrew calendar took its definitive form by integrating Meton's discovery. The Babylonians and the Chinese, for their part, also

discovered Meton's cycle and transformed their lunar calendars into lunisolar calendars.

The first solar calendars of the Maya and Egyptians

Solar calendars, which divide time according to the apparent movements of the sun (in reality the earth's movements around the sun), are necessary for agriculture, which requires that the year be divided with strict discipline into seasons. It is essential to be able to synchronize cultivation of crops with the seasons, to sow according to predictable periods of precipitation, and manage food reserves from one harvest to the next. The Maya and Egyptian calendars are the two oldest known solar calendars.

The Maya had two calendars, one religious (the Tzolkin), the other civil or political (the Haab), both based on a base 20 system, called a vigesimal system. The religious calendar had 260 days; the civil calendar (which is the one of interest to us here) had 365 days. The year was divided into 18 periods of 20 days each, and the total of these 360 days formed a *tun*. Each day had a name. Added to the *tun* were five days known as empty or phantom days, days of misfortune that were nameless. The Maya regularly kept track of the days that were lacking in a year so that the calendar remained in perfect conformity

The Jewish calendar is particularly complex. It has twelve lunar months of 29 and 30 days. Two of them vary in duration so as to adjust as well as possible to the moon. Years 3, 6, 8, 11, 14, 17, and 19 in the cycle each have a 13th month of 30 days. They are known as embolismic years. Thus the Jewish year can have six different lengths: common years last 353 days (defective year), 354 days (regular year), or 355 days (abundant year); embolismic years vary in the same manner, having 383, 384, or 385 days. Years are counted from the presumed date of the creation of the world, 3,761 years before the beginning of the Christian or Common Era. Left: a 19th-century Jewish pocket calendar. Below: a Jewish manuscript illustration depicting unleavened bread decorated for Passover.

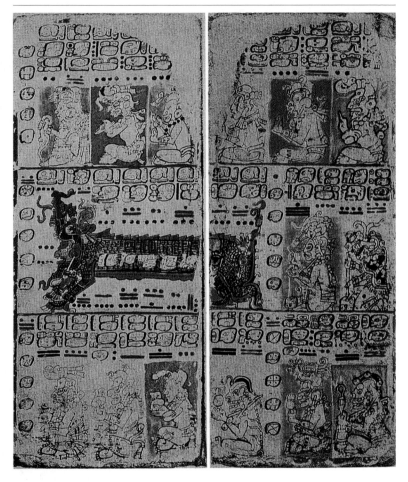

with the sun. They were able to establish an extremely precise solar calendar, despite having no reliable astronomical instruments at their disposal, by using fixed reference points.

The ancient Egyptians made great progress in articulating time. They were the first Mediterranean people to use a solar calendar, documented beginning in the 3d millennium BC. They worked it out in the simplest

The Dresden Codex, a rare Maya manuscript that survived the Spanish Conquest of Mexico, shows the extent and precision of Maya astronomical expertise.

possible way, creating a tool that was both effective and accessible. It started with the observation of a coincidence: once a year, the star Sirius rose straight up from the rising sun, and this occurred just before the annual floods of the Nile River.

For three successive years this phenomenon occurred once every 365 days; in the fourth year it happened one day later. The Egyptians divided the year into 12 months of 30 days each, to which they added five days, known as *epagomens*. They called their year "the vague year" because every fourth year the rising of Sirius occurred one day later. Since the vague year was a quarter of a day too short, it soon fell out of alignment with the seasons and did not come back into its proper phase for about 1,460 years. When the Romans conquered Egypt in the 1st century BC, they reorganized the Egyptian calendar to conform to that of Julius Caesar, but the vague year remained the reference point for the popular calendar until the 3d century AD.

Days, months, seasons

All calendars divide time into days, months, and years, but every people has its own way of experiencing these divisions and inhabiting time. The basic unit of time is the day. The alternation of day and night gives time its elementary rhythm. To compute the length of the day we use the system developed by Babylonian astronomers, who divided up the day into two sets of twelve hours, for day and for night. Why twelve? It is believed that the choice arose because of the twelve lunar months that make up a year. If the year broke down into twelve units, the day could do the same. Moreover, the Babylonians' base 60 (sexagesimal) counting system could explain their use of the figure 12, a number divisible into 60.

The day began at different hours in different calendars— at dawn in ancient Egypt and India, at sunset for the Jews, Muslims, and Chinese. The

A combination of phenomena, one heavenly (the reappearance of the star Sirius) and one earthly (the regularity of the Nile floods), served to anchor the development of the Egyptian calendar. Below: this fragment of a calendar of feast days, found at Elephantine Island, enumerates the gifts to be made to the gods on the day when Sirius rises again in the heavens, after 70 days' absence. This day had originally served to establish the date of the new year, but by the 18th Dynasty it no longer corresponded to it. Egyptian priests were vehemently opposed to any calendar reform; this variation, they said, allowed them to honor Sirius on a different date each year. Sirius is called the Dog Star, because it is the principal star of the constellation of Canis Major (the Great Dog). It rises with the sun from the end of July to the end of August, whence the name "dog days" given to this summer period.

Romans started their day at midnight, in an effort at a rational systemization. In this scheme the start of the day was equidistant from sunset and sunrise. The Roman solution has become the worldwide norm today.

In lunar and lunisolar calendars, the month is linked directly to the phases of the moon, so the length of the month depends on the moon. In India the month breaks down into two periods, a "bright" fifteen-day period from the new moon to the full moon, followed by a "dark" fifteen days. For the Muslims the month does not begin until the appearance of the first crescent, *hilal,* two days after the new moon. In solar calendars the month is dissociated from the lunar cycle. It serves as a convenient way to divide up the year.

Names of months are often inspired by particularities of the local climate or agricultural tasks. The peoples living at the mouth of the Congo River call the first

God says in the Bible: "Let there be lights in the firmament of the heaven to divide the day from the night; and let them be for signs, and for seasons, and for days and years" (Gen. 1:14). To Plato, the sun and moon are the guardians of time. He writes in the *Timaeus* (360 BC): "The sun, the moon, and the five other stars [planets] that have received the name of Wanderers appeared to define and preserve the numbers of time." Above: a miniature from a 17th-century *Book of Marvels.*

three months of the year, respectively, hunger (the sowing period), rare rains, and feminine rains (small, fine rains).

In the ancient Inuit calendar, the ethnologist Jean Malaurie reports, the names of months narrate the rhythm of life in the Arctic: the month of the moon (January), the sun appears (February), the day returns (March), the sun is warm (April), birds return (May), they lay eggs (June), the newborn birdlings fly south (August), the lakes freeze over (September), people listen (November).

Like the days, the seasons form a reality that can be perceived directly. But unlike days, seasons are not defined in universal divisions of time. Their number varies from country to country and from one climate to another. In temperate zones the year has four well-defined seasons, which generally begin with the equinoxes and solstices. In China the seasons surround these key points in time: each one begins six weeks ahead of the celestial event, a so-called period of preparation, and then "manifests itself" during the next six weeks. In tropical countries the year usually has just two seasons, the dry season and the rainy, but the popular Indian calendar lists six seasons: spring, summer, monsoon, autumn, winter, and cold season.

The year: a major unit

The largest calendar unit, the year, forms a complete cycle of the sowing, cultivation, and harvesting of plants. Each society sets a beginning for this cycle that constitutes an important threshold in collective life.

The seasons are empirical divisions of time. Their length depends on the vagaries of nature, of climate, and of vegetation. They hold an important place in traditional calendars because they regulate agricultural tasks. The Chinese distinguish four seasons. Spring begins agricultural activity and autumn closes it. In the Chinese countryside, spring was also the season of engagements and autumn the time of marriages. Above: a Chinese print depicts the activities of autumn.

In temperate countries the year starts at a pivotal moment in the solar cycle, usually near the winter solstice or the vernal (spring) equinox, more rarely around the summer solstice (as in ancient Egypt and Greece) or the autumn equinox (according to the Jewish calendar). In rural societies the year often began in spring, for instance in the Germanic, Iranian, and the first Roman calendar.

Every culture celebrated the new year with festivities, and these often assumed a special importance. People in the ancient world were less confident than we are today of the regularity of the motions of the stars. They felt the need to take part in the creation of each new year through magic practices and rituals.

Rites of renewal marked the end of the year. To ensure a new beginning, there first had to be an ending. In China the end of the year was the occasion for jousting between young and old. The elders led the year to its end and the young folk helped time to remake itself, to become young again at the start of the new year. The year sometimes ended with a festival of confusion or transgression, which provided the chance to drive out disorder before the new year. In Rome Saturnalia was a period of general license in December, a time for overturning norms, during which slaves took their masters' places. Similar exercises turn up in the Middle Ages with the Feast of Fools or the Feast of Holy Innocents.

In many societies—in Asia Minor as well as in China, Japan, and medieval Europe—the change of year occurred during a concentrated period of twelve days, which prefigured the twelve months of the year ahead. The Mesopotamians made predictions for the coming months. Farm people in France predicted the weather of each of the twelve months according to the weather of the twelve days. This twelve-day cycle, which was widespread, could correspond to the difference between the solar and lunar years. It symbolized the necessary adjustment of a traditional lunar calendar to the solar calendar, the calendar of the agricultural seasons.

Opposite: an image from the 1485 *Book of Schembart* depicts day chasing night, just as the new year drives out the old, in Carnival season celebrated at Nuremberg, Germany. A historian of religions shows the importance for traditional societies of this rupture in the calendar: "With this break in time we are witnessing not just the effective cessation of a certain temporal interval and the start of another interval, but also the abolition of the past year and of lapsed time. This is also the meaning of ritual purifications: a combustion, a cancellation of the sins and faults of the individual and of the community as a whole and not a simple purification. Regeneration, as its name indicates, is a new birth. Every new year is a reprise of time from its beginning; that is, a repetition of cosmogony. Ritual combat between two groups of figures, [the figure] of the dead, the *saturnales* and orgies are elements that denote that at the end of the year and in the expectation of the new year we have the repetition of mythic moments of passage from chaos to cosmogony."

Mircea Eliade,
The Myth of the Eternal Return, 1971

In antiquity two types of instruments served to measure time: the gnomon and the clepsydra (an hourglass using water or sand). The gnomon is the older tool. Originally just a stick driven into the ground, it gave an idea of the hour of the day and time of year, based on the direction and length of the shadow it cast. The shortest shadow was seen at noon on the summer solstice. With improvements such as lines indicating the hours, and sometimes curved lines for the days of the year, gnomons began to appear as sundials. A clypsedra is not dependent on the sun; it can indicate time intervals during cloudy weather or at night, but like a sundial it is imprecise, subject to nature's caprices. The Renaissance chronologist Joseph Justice Scaliger (1540–1609) compared sand and water clypsedras: "Those using water are… more reliable, because sand piles up or gets damp, so that it does not keep flowing. Water runs forever, wherever there is the slightest hole, but it can run out and require replenishing." Above left and right: a Ptolemaic and a 5th-century BC Greek clepsydra. Below left: a sundial and a shadow clock from ancient Egypt; below right: a gnomon from the ancient Roman city of Volubilis, in Morocco.

The week: a distinct unit

The week has a place all its own in the calendar. It is the only cycle that is entirely artificial, the only one that is not based on natural planetary movements or changes in the environment. It obeys a purely mathematical logic: the week's cycle of seven days repeats itself indefinitely. This trait leads to a second characteristic: the week is not attuned to the other divisions of time; rather, it intersects with them. The 365-day year has twelve full months but not a full number of weeks; instead, it has 52 weeks plus another day or two. We can distinguish three reasons that led certain peoples to adopt this most unusual rhythm of 52 groups (usually of 7): the need for complex divinatory cycles (as in Indonesia), the operation of a market economy (as in China and Rome), and religious motives, especially those tied to the religions of the Book—Judaism, Christianity, Islam. The Egyptians and Greeks counted on a decimal base (in units of 10). In Rome, markets were held every eighth day.

The seven-day week is a legacy from Babylonia and Judaism. It originated in Mesopotamia perhaps as long ago as the 15th century BC, where the number seven was considered unfavorable. All kinds of prohibitions hovered around the 7th, 14th, 21st, and 28th days of the month.

The Babylonian astronomers named the days of the week for the sun, moon, and the five planets known to them. The Romans later did the same, and called Monday *lunae dies,* day of the moon; Tuesday *Martis dies,* day of Mars; Wednesday *Mercurii dies*, day of Mercury; Thursday *Jovis dies,* day of Jove (Jupiter); Friday *Veneris dies,* day of Venus, Saturday *Saturni dies,* day of Saturn; and Sunday, *solis dies,* day of the sun. Languages derived from Latin still use versions of these names; English weekday names derive partly from the same source and partly from the Germanic names of equivalent Norse gods: Moon day, Tyr's (Tiw's) day, Woden's (Odin's) day, Thor's day, Frigg's day, Saturn's day, and Sun day.

The Egyptians had a tripartite concept of time. The year had three seasons: flood, sowing, and harvest; the month was divided into three periods of ten days (called decades); and the days themselves had three periods. To each category of time there corresponded a particular divinity. Below: a fragment of a 30th-Dynasty Egyptian *naos,* or box shrine, with the thirty-six decades of the year depicted on it. Each decade is represented by the image of a bird with human face standing in a boat. Each bird represents the stars in a constellation that rose at twilight during its decade; the stars were honored as gods.

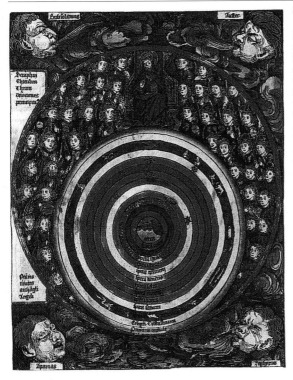

The week is very deeply embedded in biblical tradition. The repose on the seventh day is part of the Ten Commandments: "Remember the Sabbath day, to keep it holy. Six days shalt thou labor, and do all thy work; but the seventh day is the Sabbath of the Lord thy God; in it thou shalt not do any work…For in six days the Lord made heaven and earth, the sea, and all that in them is, and rested the seventh day: wherefore the Lord blessed the Sabbath day, and hallowed it" (Exod. 20:8–11). In the Jewish calendar the days are designated by their position in relation to the Sabbath: second day before Sabbath, and so on. The Christian church tried in vain to have the days of the week numbered. English has no trace of Christian terminology in its names for the days, although the other Romance languages have retained "the Lord's day" for the seventh day—*domenica* in Italian and *domingo* in Spanish, for example. Only the Portuguese language uses the Christian nomenclature for the names of weekdays. Above: a page from the 1483 *Nuremberg Chronicle* depicts the days of the week as concentric planetary rings around the earth.

The Jews were the first people to organize their life fully on the weekly cycle. To them the week was justified by Genesis: God had created the world in six days and rested on the seventh. During the Babylonian captivity in the 6th century BC, the Jews began strict observance of the Shabbat, or Sabbath—the period of rest on the seventh day—and organized their life around the rhythm of the week. Unable to pray in the Temple during their exile, they created in time what they had lost in space, giving one day out of seven, Saturday, to God. The use of the week spread thereafter throughout Asia Minor, Greece, and Alexandria to Rome. It reached India in the 5th century, and then the Far East, where it probably appeared in the 9th century. Following the

Jewish model, the other monotheistic faiths also devoted one day of the week to God: Sunday for Christians, Friday for Muslims. These groups also created a new binary rhythm, alternating regularly between rest and work, ordinary and extra-ordinary days. This rhythm proved most useful in organizing society. It is now observed all over the world.

Dating in cycles and periods

Measurement of time is not restricted to annual calendars. If it were, dates older than the current year would be impossible. Time is also counted in long-term duration, in cycles or in periods.

The earliest calendars fit into cycles. Among the Maya, the civil year had a cycle of 360 days, the *tun*. Twenty *tun* formed a *katun* (7,200 days), and 20 *katun* made up a *baktun* (144,000 days). The largest cycle included 13 *baktun*, or 5,130 Maya years. The large cycles always run in three phases: creation, use, destruction. This is common to most religions that use such a system. The Maya believed that at the end of their great cycle, which had started in 3114 BC, another cycle of equal duration would begin and a new world would be born.

The Buddhist and Hindu calendars also fit into long cycles. The Chinese system of counting time is based on

Right: a four-sided object, called a prism, from Larsa, in Mesopotamia, lists the years from 2025 to 1763 BC by the names of rulers. Such lists provided valuable chronologies for later astronomers.

a sexagesimal (base 60) cycle, which applies equally to months or to years. Each day and each year is defined by the combination of two characters—one of the "ten celestial trunks" and one of the "twelve earthly branches." The same combinations recur after sixty years.

The ancient civilizations of the Mediterranean began their dating systems from foundation events such as the start of a dynasty or the advent of a monarch. They also used brief cycles: the Olympiads (years in which the Olympic games were held) for the Greeks, the consular term of office (an administrative post) for the Romans. The major innovation in their conception of time came from Judaism. For the Jews, time has a particular philosophical meaning: the world was created by God and it will end with the coming of the Messiah. This is the first appearance of a linear concept of time, which Christianity later adopted and institutionalized. The Muslims created their own era, the *hegira,* which started with Muhammad's flight from Mecca to Medina in AD 622, known as I AH (for *anno hegira*). As these examples demonstrate, periodization depends on human rather than astronomical references. With eras, we are not only in time, but history, which allows events

Left: the great 1479 Sun Stone displays the cyclical concept of time among the Aztec. The universe, which has been created and destroyed four times, depends on the sun, whose travels across the sky are maintained through sacrifices. Thus, at the center is a fearsome sun, representing the current world. In the four squares around it are the four worlds that preceded it. Concentric circles surrounding this central theme bear the twenty signs of the sacred calendar and the different cycles of the universe.

夏至致日圖

表竿

In China astronomy played a fundamental role in the management of public affairs. Each year astronomers prepared the imperial calendar of official ceremonies. Promulgated in full pomp and sent throughout the empire, the calendar confirmed the ruler's responsibility and ability to ensure the proper conduct of the universe. Left: a Chinese astronomer measures the sun's shadow.

The traditional history of China begins with the five mythical rulers who founded the world. Below: the first was Huangdi, who

to be situated in a chronological continuity, and to be memorized.

Power, science, and religion

The calendar is thus a synthesis that draws on scientific knowledge, religious belief, and political will. It reveals the way that power, religion, and science interact. Priests carried out the functions of an astronomer in many ancient cultures, from the Babylonian to the Maya. They were responsible for keeping the festivals in harmony with the heavens, and maintaining the compatibility of religious and cosmic time.

"everywhere established the order for the sun, moon, and stars."

The calendar of Confucian China (6th century BC) gives a particularly clear example of this interpenetration. The sovereign, as son of heaven, draws his power from above and must govern in accordance with heaven. To do so he must civilize time—establish a calendar that is as exact as possible. The astronomers prepare the way and answer for their inaccuracies with their lives. According to the Sinologist Marcel Granet, the sovereign functions as an intermediary between heaven and earth. He is the master of space because he is the master of heaven: "By moving about on earth, imitating the motion of the sun, the sovereign can be considered by heaven as a son." The contact between heaven and earth is established in the Ming T'ang, the "house of the calendar," also called the "palace of light." This is a structure on a square base—representing the image of the earth—with a round roof, representing heaven.

In 6th-century BC Athens, Cleisthenes established a political calendar so that the ten tribes of the city could alternate regularly in power. In this case (as the historians Jean-Pierre Vernant and Pierre Vidal-Naquet have pointed out), the organization of time is calculated from that of space. Unlike the religious calendar, in the political calendar all periods were equivalent.

Calendars are also based on astronomy and mathematics. Astronomers measure the movements of certain stars and then arrange the days according to mathematical combinations. These sages prepare calendric systems that the political or religious authorities then enforce.

In antiquity, astronomy and astrology were thus intimately linked. No distinction of meaning was made between the motions of the stars and their influence, between observation and interpretation. Astrological readings were based on the precise and learned observations of astronomers.

Cumulative history

To improve the measurement of time it is necessary to make numerous astronomical observations and

The Chinese astronomers were elite government bureaucrats whose task was to manage an elaborate and detailed calendar. They employed many techniques in an attempt to reconcile the irreconcilable: a solar calendar composed of 24 units of 15 days each and the phases of the moon. They had to establish the length of each month such that the full moon would fall on the 15th day. Astronomers interpreted the movements of the stars, and needed to predict solar and lunar eclipses, which were understood as messages sent directly to the emperor, son of heaven. Above: the emperor celebrated solstice ceremonies and other festivals at the Temple of Heaven, built in 1420, in the Forbidden City, Beijing.

therefore to use the calendar. Through centuries and centuries of continuous observation in Mesopotamia the data were accumulated that made possible the Greek astronomy of Hipparchus (c. 190–120 BC) and Ptolemy (c. AD 85–165), who carried out precise measurements of the solar year and the lunar month.

In the 4th century BC, the conquests of Alexander the Great provoked something of a cultural revolution in the Mediterranean. The Greeks came into contact with the Persian, Mesopotamian, and Egyptian civilizations. Two calendar divisions were in circulation: the double twelve-hour day and the week. In the 1st century BC the Roman ruler Julius Caesar traveled to Egypt. His contacts with Alexandrian scholars led him to have a solar calendar adopted in Rome. Our present calendar is the product of these successive innovations: it comes from Rome, but had its origins in Mesopotamia (the 24-hour day, the god/planet names of the weekdays), Alexandria (the solar calendar), and Jerusalem (the week, the Sabbath, or Lord's day).

The collected knowledge of the ages has sometimes been lost as civilizations rise and fall, only to be recovered and reinterpreted later. In the 8th century Arab scientists and scholars rediscovered the works of Greek astronomers who had calculated the duration of the tropical year (the duration from one vernal equinox to the next) with great precision. The estimate made by Hipparchus (2d century BC) departed from reality by only five minutes.

Above: a Turkish astronomer works on a gigantic armillary sphere in this 16th-century miniature.

"Then turning his attention to the reorganization of the state, he reformed the calendar, which the negligence of the pontiffs had long since so disordered, through their privilege of adding months or days at pleasure, that the harvest festivals did not come in summer nor those of the vintage in the autumn."

Suetonius,
The Deified Julius,
from *The Lives of the Caesars,* c. AD 110

CHAPTER 2
THE JULIAN CALENDAR, FROM CAESAR TO CHRISTIANITY

In medieval calendars the religious dimension of time is less important than the agricultural. Time is measured in the farmer's labors, working the land throughout the year, with a pause in winter. Scything in July, huddling by the fire in December, part struggle and part reverie: so goes the cycle of the months. Left: the tasks of July in a 15th-century calendar; right: Christmas supper in a 12th-century sculpture.

The lunisolar calendar of the Roman Republic

To understand the organization of our current calendar, we have to go back to the calendar of the Roman Republic. In 153 BC it was established that the year began on January 1, in the period when the days had begun to lengthen. The Roman year consisted of 355 days divided into 12 months. Every two years a college of priests, the pontiffs, added an intercalar (extra) month to adjust the lunar to the solar year.

The first months of the year bore the names of gods. Ianuarius was the month of Janus, god of beginnings, who had two faces, one looking back at the year just past, the other looking forward to the new year. Februarius, month of the dead, was devoted to purifications (*februare* in Latin means "to purify"). Martius referred to the war god Mars and marked the beginning of the season of warfare. Aprilis was the month of the goddess of love, Venus, probably taken from the Etruscan (pre-Roman) name for Aphrodite, which was Apru. Maius was the month of Maia, goddess of growth, as well as the month of the *maiores,* the ancients. In Iunius people celebrated the goddess Juno and also the young. The final six months bore numbers corresponding to their place in the original Roman calendar, which had begun in March: Quintilis (fifth), Sextilis (sixth), September (seventh), and October, November, December (eighth, ninth, and tenth) months.

The months were divided into three unequal periods organized around special days that were meant to correspond to the phases of the moon: the kalends, nones, and ides. Kalends was the first day of the month and the start of the new moon. This was the day when the pontiffs announced the movable feasts of the month and when debts were paid and inscribed in the account books, or *calendaria.* Nones, the first quarter of the moon, occurred on the fifth or

"Everything you see on every side, the heaven, the sea, the clouds, the earth, everything is closed and opened by my hand. To me alone is conferred the keeping of the wide world; and the right to make it turn on its axis is mine alone."

Ovid, *Fasti,*
1st century AD

Two-faced Janus, god of doorways, ruled all transitions, both in space and in time. He ushered in the new year and the sequence of the months. Below: an ancient Roman sculpture of Janus.

seventh day; the ides, supposedly marking the full moon, fell on the 13th or 15th day. Days were counted in reverse: the first day after the ides was the 17th or 15th day before the kalends of the following month. This system remained in use until the 16th century. The word "kalends" derives from the Latin *calare,* "to call," and is the root of the word "calendar" as well as of the term "intercalary."

During favorable days commerce and other public activity was permitted; on unfavorable days it was prohibited. Public festivals, or *festi,* fell on unfavorable days, reserved for religious observance.

Julius Caesar took the title of pontifex maximus in 63 BC and dictator in 49 BC, acquiring unprecedented political and religious powers. By this time the calendar was in

This 3d-century AD mosaic combines profane and sacred depictions of time. The seasons are represented at left by figures with rural occupations; the twelve months are mostly symbolized by ceremonies of the pagan cults: Lupercalia in February, Saturnalia in December, and so on.

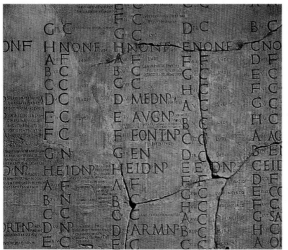

The pre-Christian Roman calendar was established by the Pontifex Maximus, the highest religious authority of Rome, and functioned as a registry of civil time. For each month it indicated market days with letters A through H, favorable days (*fasti*) with an F, unfavorable days (*nefasti*) with an N, religious feasts (*festi* and *feriae*) with NP, and dates of popular assemblies (*comitiales*) with a C. In reforming the calendar, Julius Caesar took care not to change the sequence of ceremonies, but modified the spirit of the calendar by introducing new holidays to commemorate his own victories. Thus was the cult of the emperor born. Left: fragment of a Roman calendar. Below: a portrait of Julius Caesar.

serious disarray. The pontiffs had taken liberties with the intercalar months, and the legal year was three months out of alignment with the seasons. In 46 BC the ruler decided to reform time. He wanted to establish his dominion on a durable foundation, correct old errors, and build a solid empire. Calendar reform would unify all the conquered peoples of the expanding Roman Empire under a single system of measuring time.

The Greek astronomers of Alexandria, in Egypt, were the supreme scientific authorities of that time. On their recommendation Caesar established a solar calendar. Since the duration of the solar year was then estimated at 365¼ days, Caesar introduced a four-year cycle: the first three years were 365 days long and the fourth 366 days. The extra day was added to February by doubling 24 February, the sixth day before the kalends of March. This fourth year was called *bis sextus* ("double six"); it corresponds to our leap-year day. The months alternate between 30 and 31 days in length, except February, which has 28 or 29.

Caesar added favorable days to stimulate commercial enterprise.

Finally Rome had an effective instrument for managing its dates. Even if the new calendar lacked precision—the solar year was slightly shorter than the assumed estimate—it was clear, accessible, and orderly. It made the planning of future activities much easier, and thus served the needs of the state and the army greatly. The pontiffs could no longer add extra months at will, and their power was much reduced. The calendar lost some of its references: months no longer depended on the lunar cycle, so that the kalends, ides, and nones became purely conventional divisions.

With this reform Julius Caesar installed a solid temporal architecture. The Julian calendar remained in force for one and one-half millennia, until 1582.

Christianity invests the calendar

In the late Roman Empire Christianity spread swiftly. By the 4th century Christians made up a considerable portion of its population. With Emperor Constantine's conversion to Christianity in 313, they began to assume a privileged status and soon succeeded in having their organization of time accepted by the Empire as a whole. At that time the expanding Christian church appropriated many Roman imperial institutions, including the Julian calendar, which gradually became Christianized. The church grafted its liturgical dates onto the calendar, establishing its own cycle of ceremonial and commemorative days and replacing the pagan Roman—an enormous task of acculturation.

Built around the two major holy days of Christmas and Easter, the Christian

In 44 BC the fifth month, Quintilis, was renamed Julius (July) in honor of Julius Caesar. In 8 BC Sextilis became Augustus (August), to glorify the Emperor Augustus, portrayed below in a bronze statue.

calendar recalled the life of Christ: nativity, death, resurrection, ascension. It was fine-tuned between the 4th and 9th centuries and has scarcely changed since.

Christians strove to set themselves apart from their forebears the Jews and sought to distinguish their calendar from the Jewish one. They retained the seven-day week and the idea of devoting one day out of each weekly cycle to God, but after lengthy controversies, they chose the day after the Jewish Shabbat, since Christ's resurrection had taken place on that day. Thus, because Judaism kept the Lord's day on Saturday, Christianity associated itself with Sunday. Gradually the Roman Empire adopted the rhythm of the week and the idea of a day of rest, devoted entirely to religion, to be spent as a community, in church. In 321, Constantine prohibited public activities in cities on Sundays.

Christmas and Easter

The dating of Easter has prompted numerous debates. According to the Gospels, Christ died on the first day of Jewish Passover. According to the Hebrew calendar this is the day of the full moon that follows the beginning of spring. It is therefore a movable feast, changing its date every year. At first Christians tried to find a date for Easter that would not depend on the Jewish calendar but that would still be at the beginning of spring. At the Council of Nicaea in 325, it was decided that the festival of Easter should be observed on the same Sunday by all Christians, on the day of the full moon that follows the vernal (spring) equinox. Unfortunately, for some time there was no widely accepted and accurate means forecasting the dates of movable feast days, but in the 8th century the vernal equinox was fixed on March 21, by means of the Nicaean computation method.

The decision had great repercussions. Because Easter was a movable feast, the Christian calendar became highly complex, embracing both a solar system (organization of the year) and a lunar (organization of the pascal or Easter cycle). The ability to forecast the

Easter, the climax of the Christian calendar, each year recalls the death and resurrection of Christ as the sacrifice made to save humanity. Above: a detail of a Byzantine mosaic of the Crucifixion. The entire calendrical cycle from Carnival to Pentecost depends on Easter (celebrated between March 22 and April 25), whose mobility emphasizes both its importance and its mystery, since the date of its observance is constantly shifting. Today the movable date of Easter introduces an element of fantasy and irregularity in an otherwise predictable calendar.

Left: a Byzantine mosaic of the Nativity. The feast of Christmas was instituted by the Christianized Roman Empire to combat pagan celebrations of the winter solstice. It also sought to demonstrate the divine nature of Jesus at a time when the Arian heresy (a doctrine holding that Christ was a mere mortal and not the son of God) was spreading confusion among Christians. In the East, where Arianism was popular, the church celebrated Epiphany (January 6) as the manifestation of God on earth for the same reason. Christmas was soon inscribed within a cycle of transgressive celebrations that lasted until Epiphany: the Feast of Fools, Holy Innocents' Day, the Feast of the Donkey—these were names given to satirical, licentious demonstrations in which the hierarchy of church authority was reversed. "For several centuries the church of Constantinople permitted the people and the clergy, on the feasts of Christmas and Epiphany, to indulge in shouting, noise, dancing, and buffoonery in the middle of the temple and the sanctuary," Patriarch Theophylactus lamented in the 10th century. These "nonsense games" were eventually prohibited by the church and had mostly disappeared by the late Middle Ages.

date of Easter each year depended upon solid knowledge of astronomy. In fact, attempts to improve the accuracy of this computation were a matter of constant scientific speculation throughout the Middle Ages.

The convention by which Christ's birth was celebrated on December 25 also began in the 4th century. The Gospels give no information about Jesus's birth date. Why was that day chosen? There seem to be two complementary explanations for the choice.

The early theologians and scholars of the church

wished to anchor the principal Christian feast days to certain key points in the solar year: Easter was linked to the vernal equinox, Christmas close to the winter solstice, when the days start to lengthen, with their promise of the renewal of spring. This pivotal point in the year had also given rise to the extremely popular pagan holidays: December 24 marked the end of Saturnalia, the long Roman festival that was the origin of Carnival. The next day, December 25, was the celebration of the birth of the unconquered sun (*natalis solis invicti*). This was a ceremony in honor of Mithra, an Asiatic god of light, whose cult, imported from the Near East by Roman soldiers, was officially recognized by Emperor Aurelian in AD 274. Christmas was superimposed on Mithra's celebration. In one of his sermons, St. Augustine (AD 354–430) calls Christ the sun of justice. Under pressure from Christian bishops and emperors, Christmas supplanted the pagan cults of the solstice.

The calendar speaks with a Christian voice

The Christian adaptation of the Roman calendar was so successful that it was accepted throughout the Western world, even in non-Christian contexts, and influenced many secular governments and institutions. Liturgical time is a highly elaborated cyclical time. The Christian liturgical year starts in late November with Advent, a time of penitence that precedes Christmas, interrupted by a few major feasts

Above: the Salzburg Calendar, drawn in AD 818, fixed the great medieval themes representing the months. These were organized around the activities of rural and agricultural life. Hardship is not shown; rather, the emphasis is on production and abundance, especially the cultivation of wheat for bread, grapes for wine, and livestock. Leisure is not forgotten: people are shown at hearthside pastimes and celebrations. Many calendars of this type, painted or sculpted, are found in churches. Left: the column of Souvigny, sculpted in the 12th century. How are we to interpret these profane depictions placed in a religious context? Is this a pedagogical effort, an attempt to give solid content to the twelve months, to make the division of time into set units accessible to everyone? Was it a religious lesson—after the Fall, humankind is redeemed by work? These calendars sometimes retained pagan memories. Right: December is represented by the Roman holiday of Saturnalia on the Door of the Months of the cathedral of Ferrara.

such as that of St. Nicholas (December 6), which remains very popular in Central Europe. Christmas is followed by 12 uninterrupted feast days, ending with Epiphany on January 6. Then comes Carnival, which is the period before Lent, and before Easter, and therefore a movable feast. It often begins on Candlemas (February 2). It is traditionally a time of freedom, release, and transgression, and its festivals admit all sorts of excessive behavior. Lent ends it with 40 days of penitence (excluding Sundays), from Ash Wednesday to Easter. Agricultural work resumes during this period. The two months that follow Easter are marked by major holy days: Rogation Days, Ascension (40 days after Easter), and Pentecost (10 days later), but these are periodic events, competing with the agrarian rites that celebrate the renewal of nature and the return of fair weather: the maypole of May Day (May 1), bonfires on St. John's Day (June 24).

The liturgical and agrarian calendars complement one another. Agricultural activity is intense from May to October. The high points of the liturgical calendar take place during the agricultural dead season, from November to Easter. The whole ensemble forms a coherent structure, perfectly adapted to a rural society.

The church also dominates the weekly cycle. Sunday enjoys a status completely distinct from the other days of the week. Sunday allows the church to affirm its mastery over society. According to the historian Jacques Le Goff, "As the Lord's day, the day of obligatory mass for everyone, celebrated by priests, Sunday enabled the church to exercise regular control over economic and social time."

By arranging the calendar, Christianity imposed a way of conceiving of time on society

"Time is marked with religious points of reference whose meaning has become inseparable from the time of year in which they occur. All Saints' Day and All Souls' Day are associated with autumn and the beginning of the earth's sleep;…Easter and the Resurrection with spring and the renewal of nature."

François Lebrun,
Croyances et cultures dans la France d'Ancien Régime (Beliefs and Cultures in Pre-Revolution France),
2001

In the foreground of this 1559 painting by Pieter Brueghel the Elder, *The Combat of Carnival and Lent,* Goodfellow Carnival, astride a keg, and with a spitted suckling pig for a lance, confronts Old Lady Lent, with her loaves and fishes, thus marking the break in the calendar that indicates the end of winter and separates "fat" days from "lean," days of folly from days of repentance. Carnival and Lent appear to joust, but they are not truly fighting: Carnival waves goodbye, while Lent awaits her approaching hour, with the ashes of Ash Wednesday already marking her forehead. The two processions bypass one another, for Carnival and Lent are side-by-side in the Christian calendar. The first is the indispensable prologue to the second. On the eve of the forty days of Lenten penitence, villagers and townsfolk "bury" their pagan life by performing rites of the ancient winter holidays—role reversals and masquerades, to which Christianity gives a new meaning. "Carnival incarnates unbridled pleasure, delight in living and dancing, the gross pagan sins that will be expelled excrementally by Lent," writes the historian Emmanuel LeRoy Ladurie. If Lent were to relent, Carnival would weaken.

This illumination from the *Psalter of Blanche of Castile* depicts the masters of the calendar: the astronomer who holds an astrolabe, the chronicler who presents his work, the copyist who records events. In the Middle Ages knowledge of time was the preserve of a very small number of people, educated members of the church. They alone had a quantitative idea of the calendar, since the days all had the same value in their calculations. Astronomers enjoyed considerable prestige: the ability to read the heavens, to measure and master time, put one close to God. Kings and popes each had a court astronomer. The 13th century saw a great revival of mathematical and astronomical knowledge, thanks to translations of ancient Greek and Arabic texts. Knowledge of the heavens progressed rapidly. Yet the church alone measured time. And the church showed a scientist's concern for precision in observations and calculations. Right: this medieval angel on the facade of the cathedral of Chartres probably acquired its sundial in about 1538.

as a whole. This shows up clearly in Christian discipline. The liturgical calendar includes periods of penitence, of fasting, and of sexual abstinence. Christians could not have sexual relations during Advent or Lent, on Sundays, holy days or the day preceding them, or at the start of each season, during what are called Ember Days.

The Christian calendar also structures economic life. The church prohibits working on holy days. The great fairs—vital economic engines of commerce in the Middle Ages—were held during periods of fasting: the fair of Lagny occurred in January, during the 12 days after Christmas; that of Provins on the Feast of the Holy Cross, May 3. Debts were paid in autumn, usually on St. Michael's Day (September 29) but sometimes on St. Remy's (October 1) or St. Martin's (November 11).

The church also regulates the measurement of time. Only clerics were permitted to calculate the date of Easter and, consequently, the date of all the ceremonies that depend on Easter, from Mardi Gras to Pentecost. Clerics alone, therefore, established the calendar. Tradition held that parish priests announced the date of Easter from the pulpit on Epiphany (January 6), thus demonstrating the church's grip on the calendar.

Saints and angels

Aside from the liturgical cycle the church established another religious cycle, that of the saints' days. From the beginnings of Christianity the apostles, martyrs, and confessors had inspired their own individual devotions. The saints and the Virgin Mary were considered intercessors with God and their placement on the calendar was no more left to chance than had been the feasts of Christmas and Easter. The church fixed the feasts of the major saints at pivotal moments in the solar year—periods of pagan holidays it wished to supersede or integrate into the Christian world view. "Not just any saint could replace any god," the ethnologist Claude Gaignebet points out. Thus was born the calendar of saints.

The feast of St. John the Baptist falls on June 24, quite close to the summer solstice, and is exactly symmetrical with the Nativity of Jesus. This arrangement may have been inspired by a passage in the Gospel of St. John: "He must grow as I diminish." St. Augustine explains it thus: "For it is with the birthday of John that the days start to grow shorter, but the birth of Christ starts the renewal of their growth."

The archangels are also celebrated on significant dates. With their days set at the equinoxes, archangels Gabriel (March 25) and Michael (September 29) watch over the conception of Jesus and of John the Baptist, respectively, whose births coincide with the solstices.

The period separating Christmas from Epiphany forms a cycle of 12 feast days. Major saints' days occur during this period: St. Stephen, the first martyr, on December 26; St. John the Evangelist on December 27; Holy Innocents, patron saints of choirboys, on December 28. All this is no doubt intended to give a Christian meaning to the end-of-year festivities.

In February, wild animals such as wolves and bears were believed to emerge from the forest so that winter could end. St. Blaise is celebrated on February 3, since legend has it that he could speak to wild animals.

A very old pre-Christian tradition called for bonfires to be lighted in the beginning of May. The church instituted the feast of the finding of the True Cross on May 3. Rogation Days, celebrated for the first time in 469, marked another attempt to Christianize agrarian rites of May. For three days before Ascension Day the priest and villagers paraded through the fields of the village to ensure good crops.

Another significant example of Christian acculturation is seen in All Saints' Day, the holy day for all the saints who have no place of their own in the calendar. It is an old tradition in the Eastern church and was known as early as the 4th century in the West, where it originally

St. Michael is the warrior archangel who vanquished Satan and who weighs souls at the Last Judgment. He is thus also the messenger of the dead. His day is September 29, a date established by Charlemagne. St. Michael's Day was a high point in the Christian religious and economic calendar; seigneurial taxes were paid on that day. Above: a 15th-century illumination of the archangel.

occurred in the spring. Pope Gregory III (731–41) moved it to November 1, the date of Samhain, the very popular Celtic day of the dead. Two centuries later, All Saints' had become one of the major Christian holy days. (The night before it, All Hallows' Eve, eventually became the carnivalesque holiday of Halloween.) In the 10th century a priest named Odelon of Cluny instituted the Day of the Dead on November 2. And so Samhain was completely assimilated.

Thus, by placing its feast days judiciously, the church managed to Christianize all the key periods of the year. The period of Christian feast days culminated in the 13th century with the creation of the Feast of Corpus Christi, 10 days after Pentecost, and the development of the cult of the Virgin, with her special days: the Visitation, Assumption, and Immaculate Conception.

Julius Caesar had politicized the Roman calendar. First he, and then his successor, Augustus, had used it to promote an imperial cult. Christianity appropriated the calendar and modified it, elaborating a ritual that banished the scientific measurement of time under the symbolic weight of Christianity.

Toward a linear conception of time: the Christian era

An even more radical change in perspective had occurred at the dawn of the Christian era. The date of the year

The feast of St. John the Baptist, June 24, casts a thin Christian veil over the summer solstice. Above: a 15th-century image of the saint. On the summer solstice, a privileged moment of contact with cosmic forces, bonfires were lit and herbs collected by night. Below: the blessing of the wheat in Ardois, France, in the 19th century.

f			Decemb.
g	iiii	n	
A	iii	o	
b	ii	n	Barbare ṽ.
c		no n̄.	
d	viii	i	Nicolai epi.
e	vii	d	
f	vi	v	
g	v	f.	
A	iiii	i	Abundii mr.
b	iii	d	
c	ii	vf.	
d		idus f.	Lucie ṽ. ⁊ mr.
e	xviiii	k Januarii.	
f	xviii	f.	
g	xvii	k	Ananie. dʒaē. ↄcē.
A	xvi	f.	
b	xv	k	
c	xiiii	f.	
d	xiiii	k Vigilia.	
e	xii	p thome apli.	
f	xi	k	
g	x	f.	
A	viiii	k Vigilia.	
b	viii	p natiū dn̄i.	
c	vii	k Stephani pʒ̄.	
d	vi	k Johannis ev̄.	
e	v	k Innocentum.	
f	iiii	l.	
g	iii		
A	ii	k Siluestri ẽ ⁊ mr.	

Medieval calendars often took the form of sculptures on church doors, but in the late 13th century they began to turn up in books of hours, expensive hand-painted breviaries meant for well-to-do individuals. These calendars served to orient the user within sacred time; they had less to do with measuring time than with establishing values. Days were not numbered, but were identified with liturgical festivals or the commemoration of saints. To find the connection between a date and a day of the week, one had to know the dominical letter assigned to the year, which indicated the place of the first Sunday of January in the week (January 1 was "A," January 2 was "B," and so on through the number 7, corresponding to "G"). Far left: in the 13th-century *Psalter of St. Elizabeth* December is represented by the annual butchering of pigs, St. Nicholas's Day, and Christmas. Near left: in the 16th-century book of hours of Anne of Brittany the fattening of pigs, preamble to their slaughter, illustrates the first half of November.

| | | | | Septembre a . xxx . iours. | la quātite. | le nôbre |
| | | | | Et la lune . xxx . | des iours | dor. |
					pore minuit.	nouuel.
rbi	f			sainct leu. sainct Gille.	xj.	xlij.	
v.	g	iiij	f̊	sainct antoyne.	xij	xlix	ij
	A	iij	f̊	sainct godegram.	xij	xlvj	x.
xiij.	b	ij.	f̊	sainct mamel.	xij	xliij	
ij.	c	Nonas	sainct macouin.	xij	xl.	xviij
	d	viij	id̊	sainct donacen.	xij	xxxvj	
x.	e	vij	id̊	sainct clouet.	xij	xxxiij	vij.
	f	vi.	id̊	Noshedame.	xij	xxxvij	xv.
xviij	g	v.	id̊	sainct omer.	xij	xxiiij	iij
vij	A	iiij	id̊	sainct gobert.	xij	xx.	
	b	iij	id̊	sainct prothin.	xij	xvi.	xij
xv:	c	ij	id̊	sainct laur.	xij	xij	
iiij.	d	Idus.	sainct regnalt.	xij	vij.	i.
	e	xviij	kl̊	saincte aioys.	xij	iiij	ix
xij	f	xvij	kl̊	sainct nichomede.	xij	v	
i.	g	xvi.	kl̊	sainct euferinne.	xi.	lvij	xvij
	A	xv.	kl̊	sainct lambert.	xi.	liij	vi.
ix.	b	xiiij	kl̊	sainct fenol	xi.	l.	
	c	xiij	kl̊	sainct signe.	xi.	xlvj	xiiij
xvij	d	xij	kl̊	vigille.	xi.	xliij	
vj.	e	xi.	kl̊	saincte mathieu.	xi.	xxxix	iij.
	f	x.	kl̊	sainct morice.	xi	xxxvj	xi.
iiij	g	ix.	kl̊	saincte egle	xi.	xxxiij	
iij.	A	viij	kl̊	sainct lier.	xi.	xxix	xix
	b	vij	kl̊	sainct firmin.	xi	xxvi.	vij
xi.	c	vi.	kl̊	sainct aprien.	xi	xxij.	xvi
xix	d	v.	kl̊	sainct cosme.	xi.	xix	
	e	iiij	kl̊	sainct prefine.	xi.	xv.	v.
viij	f	iij	kl̊	sainct michiel.	xi	xij	
	g	ij.	kl̊	Sainct gerosine.	xi.	ix.	xiij.

Les Très riches heures du Duc de Berry (*The Very Rich Hours of the Duke of Berry*), painted in the early 15th century, offers an elegant calendar of the late Middle Ages. Far left: the actual calendar portion provided a guide to sacred time. In columns from left to right are: the golden number of the cycle of Meton, the dominical letter (from A to G, to identify the place of Sundays in the year), the Roman numbering of the days (kalends, nones, ides), the cycle of saints' days, and the length of each day. Near left: at the top of each illustration, in the vault of the heavens, the days are numbered in an attempt at a scientific measurement of time. Here, the depiction of September reflects a strictly hierarchical conception of the universe. In the foreground of the illustration the abundance of the earth is harvested; in the background are the royal splendors of the Chateau of Saumur, a flamboyant palace inaccessible to the peasantry. At top is the dome of heaven, with its ancient symbolism: the chariot of the sun-god Apollo and the signs of the zodiac through which the sun is passing in September.

was no longer linked to the reigns of kings or dynasties, as it had been in antiquity. Instead, the years began their numerical sequence with the birth of Christ. This fixed date as the point of departure for the calendar changed the perception of time.

Dating by year was not unknown before Christianity, but it took many forms, most of them quite local. A year might be counted as so many years *ab urbe condita* (abbreviated as AUC), that is, after the founding of Rome. This was understood to have occurred in the year we now call 753 BC. In Asia Minor dates were set from the foundation of the Seleucid dynasty (312 BC); in Spain from the Spanish era, 38 BC, to cite just the most widespread systems. In the West, people also counted from the years of the reign of Roman emperors or attempted to count the years since the creation of the world, AM, or *annus mundi.* Every chronologer had tried to estimate the date, with widely varying results. (One of the most common was that of the Byzantines, 5508 BC.)

One system proved more useful: indiction, the position of a year within a 15-year cycle. First used for income taxes, indiction was a cyclical year-dating method reserved for public administrators and used continuously until the 12th century. The West thus employed various dating systems, most of them cyclical and pagan. The Christian era resulted from the desire of chronologers to fix the date of Easter precisely. The first Pascal (Easter) tables came from recently Christianized

The Christian dating system developed by the monk Dionysius Exiguus (Denis the Short) was spread through the West and eventually adopted thanks to the work of an English monk named the Venerable Bede. In 725 this eminent early-medieval expert on time published a treatise, *De temporum ratione (On the Reckoning of Time)*, known throughout Christendom, in which he supported Dionysius's thesis. He then wrote his masterful *Ecclesiastical History of the People of Britain*, whose chronology was based on the Christian era as established by Dionysius. In *De temporum ratione*, Bede established tables to determine the date of Easter up to the year 1063, based on a cycle of 532 years (the 19 years of Meton's lunar cycle multiplied by the 28 years of the solar cycle, after which a date recurred on the same day of the week). These tables were accompanied by a manual for the use of chroniclers, which remained popular throughout the Middle Ages. Left: a page from Bede's tables.

Left: this medieval depiction of the solar year is based on concepts from Mesopotamia. By about 600 BC Babylonian astronomers had identified the ecliptic, the sun's apparent course around the earth, and divided it into 12 parts, each named for a constellation in which the sun rose during that period. The Greeks adopted this division of the heavens and of time, calling this set of 12 calendrical constellations the zodiac (*zodion* means "animal figure" in Greek). In the Middle Ages the zodiac remained a full-fledged part of the Christian world view, despite its pagan roots. Sculpted representations of it often are found in medieval churches, especially on the facades, in books of hours, and in manuals for the reckoning of Easter.

Egyptian Alexandria and took the reign of Emperor Diocletian as their point of departure (AD 284), because in that year the lunar Epact—that is, the number of days elapsed since the new moon that precedes the beginning of the year—was zero, as it was supposed to have been at the time of the creation of the earth (or so the chronologers believed). The idea of the Christian era came from the rediscovery of these tables. In the mid-6th century Pope John I appointed a Scythian monk, Dionysius Exiguus (Denis the Short) to update the 5th-century Pascal tables of Cyril of Alexandria, so that Rome would have its own computation system and no longer depend on Alexandria. Dionysius elected to date his tables, not from the reign of Diocletian, who had been a terrible persecutor of Christians, but from the Incarnation of Christ, "the better to make known our hopes and to celebrate the redemption of humanity."

Dionysius placed the birth of Jesus on December 25, 531 years before his own time. How he arrived at this number is not clear. He named the point of departure of his dating system *anno domini,* in the year of our Lord, 1, abbreviated as AD 1.

Today we consider this established date as wrong in two respects. In his Gospel, St. Matthew says that Jesus was born in the reign of King Herod, who died in 4 BC. Dionysius must have been wrong by at least four years, probably five or six, according to specialists. Considering this error in dating the birth of Christ, the literal year 2000 of Christianity occurred several years earlier than the millennial date celebrated recently.

Moreover, Dionysius started dating directly from year 1 because in his time the West did not yet have the concept of zero. This is mathematically incorrect, since it does not account for the year between 0 and 1. Since the early 18th century, this question has inspired recurring debate: should we make centuries begin in the year 0, a round number in our decimal system, or in 1, so that each century truly lasts 100 years? When do we mark the 1st millennium, in the year 1000 or in 1001, and when did we enter the 3d millennium, in the year 2000 or in 2001?

The use of the *anno domini* dating system spread in the 10th century, when the habit of dating official documents began to be popular. The Christian era joined other dating systems, such as indiction and *annus mundi*. When Pope Boniface VIII announced a Jubilee for the year 1300 and large crowds converged on Rome, it was

For centuries theologians, astrologers, and mathematicians have tirelessly examined mentions of numbers in the Bible in an attempt to calculate when time began and when it would end. Until the 18th century the most widespread belief claimed that the world would last 6,000 years. The thinking was this: God had created the world in six days and biblical passages assigned a value of 1,000 years to each of God's days: "For a thousand years in thy sight are but as yesterday when it is past" (Psalm 90); "One day is with the Lord as a thousand years and a thousand years as one day " (2 Peter 3:8). From this, people concluded that the history of the world would comprise 6,000 years. There remained one crucial question: the date of the creation of the universe. (left: a medieval representation). The Septuagint Bible had given the Creation a date of 5500 BC, but Bede calculated it at 3,952 years before his own time. In 1650 James Ussher, the Anglican bishop of Ireland, announced that the universe had begun on October 23, 4004 BC at 9 o'clock in the morning, a date validated by Isaac Newton.

clear that Dionysius's system had become well implanted in the consciousness of Europe. At first this dating method was used for events later than the birth of Christ, but starting in the 17th century it became the reference point for dating events from pre-Christian antiquity as well. Counting proceeded backward from the presumed birth date of Christ, to form the dating system called BC, that is, before Christ. The counting of years from the birth of Christ remains the most common system worldwide, although other systems, both religious and secular continue to be used. Modern dating systems sometimes rename BC and AD as BCE (before the Common Era) and CE (Common Era).

The church was mainly concerned with the date of Easter. It embraced the concept of Christian era and accepted this dating method, but never legislated it officially.

Above: detail of Giotto, *The Last Judgment*, 1306.

"The sun shall be darkened, and the moon shall not give her light,…and the powers that are in heaven shall be shaken. And then shall they see the Son of man coming in the clouds with great power and glory. And then shall he send his angels, and shall gather together his elect from the four winds, from the uttermost part of the earth to the uttermost part of heaven."
Mark 13:24–27

IV. Saint agath

The Christian era represents a rupture in the conception of time: the incarnation of Christ occurred on a specific date in history, a precise point in time, and this fact gave new meaning to what followed. Time was ordered along an axis, with a central point, the birth of Jesus, a before and an after. For Christians, history is in motion in one direction, from a fixed beginning toward an understood end, the Last Judgment. Christian time lost the sense of cyclical and recurrent eras and became linear, unidirectional, and irreversible. The universal symbol of the circle that had characterized the eternal return of days, months, and years in ancient religions was replaced by the arrow. This linear view of time soon began to infiltrate the collective consciousness; it created historical perspectives and served to distinguish clearly the past, present, and future as discrete concepts. Humans began to situate themselves with precision in the succession of generations.

A kaleidoscope of times

Christianity infused the calendar with its particular meanings, and the Christian era was adopted throughout the West. But this did not mean that the Christian calendar in the Middle Ages was unified. On the contrary, diversity ruled. Saints' days were not celebrated on uniform dates, but followed local and regional traditions. Days started at diverse times—dawn, sunset, or midnight. The year did not start everywhere on the same date. According to one system the year began on the traditional date of Jesus's circumcision, January 1; another system began the new year on the nativity, December 25; still another on March 25, the date of the Annunciation, which is also the supposed date of the creation of the world. The Eastern system, which started the year with Easter, was the most widely followed as well as the most difficult, since Easter is a movable feast, so that every year began on a different date. In Byzantium the year began on September 1, in Russia on March 21, with the vernal equinox. New Year's day even varied from one

Left: a line referring to Saint Agatha's Day from a book of hours.

"The Christian faith is based on the idea that the life of Christ cuts history in half. Thus it has an essential sensitivity to chronology. But this chronology is not based on a concept of time as divisible into equal, precisely measurable instants—what we call objective or scientific time. It is a chronology imbued with significance. The Middle Ages, as eager as we to assign dates to events, nevertheless did not date according to the same norms and the same needs…Everything concerning Christ is marked by a need for temporal measurement."
Jacques Le Goff,
La Civilisation de l'Occident médiéval,
(*The Civilization of the Medieval West*), 1964

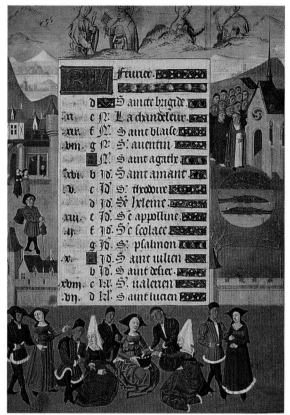

Books of hours were of no use in measuring time, but they helped people to find their place in the year and to become familiar with the sequence of the months. Unlike the seasons and days (linked to solar time), or the week (linked to the liturgy), months corresponded to nothing concrete. They formed a division of time that was completely abstract. They therefore had to be given meanings, characterized by activities appropriate to their seasons. In a book of hours painted around 1460 for the Duchess of Burgundy the months are associated with games and festivals. Left: the first half of February is illustrated by a game of charades, using little slips of paper, and by Candlemas, the feast of the purification of the Virgin and the presentation of Jesus in the Temple, on February 2 (which is also Groundhog Day in the United States). Oral calendars of sayings and proverbs also helped to orient people in time: "If Candlemas day be fair and bright, / Winter will have another flight. / If Candlemas day be shower and rain, / Winter is gone and will not come again."

city to another. In Venice it was March 1; in Reims March 25; in Soissons in the 13th century it was December 25. In addition, monarchs and popes imposed their own systems. In France New Year's day changed repeatedly. In 1564, when Charles IX fixed it for the entire kingdom of France on January 1, he met with stiff resistance from the provinces, which held to their particular calendars. The Paris parliament only applied his edict in 1567, three years after it was promulgated. This patchwork of time-frames was habitual and widely accepted, but as the modern nation-states evolved they grew eager to standardize their administrations, and it became ever more problematic.

Left: the page for July and August from an 11th-century ecclesiastical calendar.

In the Middle Ages the calendar imposed a rhythm on time. By the end of the 16th century it was becoming a more sophisticated device: a means of defining past and future. In the Renaissance the civil calendar began to free itself from the religious year.

CHAPTER 3

AN INSTRUMENT OF MEASUREMENT

In the late 13th century the mechanical clock was born from the demand for precision in time measurement. Its invention unleashed a veritable revolution in culture. Left: a 14th-century astronomical clock in Lund, Sweden. Right: a French Revolutionary calendar for the Year II bears the mottos, "Unity, Indivisibility of the Republic, Liberty, Equality, Brotherhood, or Death."

In this 16th-century manuscript illumination astronomers in Constantinople study the heavens using a number of observational and computational tools. At center a scholar in blue holds a quadrant (the instrument shaped like a quarter-circle), which measures the angle of elevation of a star, and thus one's latitude on earth. Taking successive measurements of a star's position relative to earth is a means of determining the passage of time. Behind him another scientist holds an astrolabe, the calculation instrument indispensable to astronomers, which measures the altitude of a star such as the sun, and from that determines the hour. These instruments were developed and refined over many centuries by many peoples, but it was the Arabs especially who inherited and preserved early Greek astronomy, with its collected data about the positions of the stars over many years. Starting in the 8th century, under the Abbasid dynasty of Baghdad, Arab and Persian scholars translated ancient Greek and Indian scientific texts. They mapped the heavens and drew up astronomical charts to calculate the planets' positions at various dates. The Ottoman Turks in turn benefited from this science in the 16th century.

A more precise sense of time

In the 12th century the mathematical knowledge of the Arabs began to spread through North Africa and Spain into Europe. Treatises translated the discoveries of Arab scientists into Latin and described the numerical system developed by Indian mathematicians and refined by the Arabs: positional notation, the use of nine numerals plus zero, and decimal fractions. These innovations permitted faster and much more advanced computation than the old Roman numerals and proved enormously helpful to astronomers. Al-Battani (c. 859–929) in Baghdad and later Omar Khayyam (1048?–1131) in Persia managed to quantify the duration of the solar year with near exactitude. Astronomical charts developed by the mathematician al-Khwarizmi (c. 780–c. 850) were translated into Latin in 1126. In about 1252 the so-called Alfonsine tables were compiled by astronomers working for King Alfonso X of Castile. These were a collection of highly reliable data about the positions and movements of the planets that became the standard reference for European astronomers until the end of the 16th century.

Paradoxically, the fascination with astrology in the 14th and 15th centuries encouraged progress in time measurement. This new interest was probably a reaction to a widespread sense of insecurity in Europe, fostered by plague epidemics, wars, and the growing crisis of the Protestant Reformation. Astrology reflects a desire to foresee the future, to gain some idea of what to expect. It motivated scientists to seek better knowledge of the movements of the celestial luminaries, that is, of the sun, moon, and planets. "If the confident hope of reading the future in the heavens did not exist, would people be wise enough to study astronomy for its own sake?" the German astronomer Johannes Kepler (1571–1630) wondered in the late 16th century. Both he and the Danish astronomer Tycho Brahe (1546–1601) apparently practiced astrology with pleasure.

A new way of experiencing time

The economic prosperity of the West in the 12th and 13th centuries brought about a transformation in people's relationship to time. European cities

The astrolabe was known in antiquity and was thought to have been invented by Hipparchus in the 2d century BC. The instrument can calculate the position of a star at any time during the year and at any latitude and was vital to sea navigation. It was used first in Byzantium and Baghdad and then introduced into Europe by Jewish scholars in Arabic Spain in the 10th century. Its use spread over the next two centuries with the translation of Arabic texts that explained its use. Below: a 17th-century Arab astrolabe.

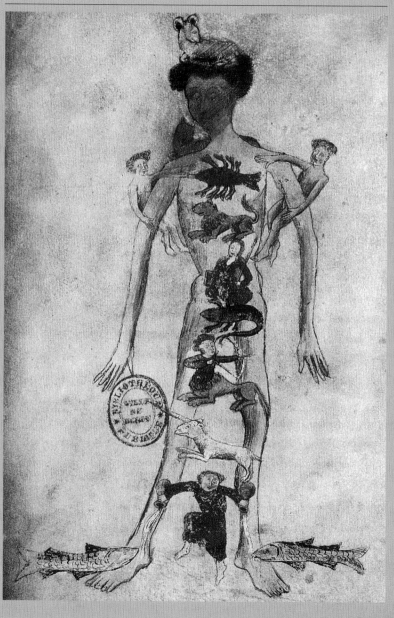

Until the 18th century astrology was considered a scientific discipline on an equal footing with astronomy. Shaped by the Alexandrian Greek scientist Ptolemy in the text called the *Tetrabible*, written in the 2d century AD, astrology was widely practiced in Byzantium and in the Arab world. It was rediscovered by the Christian West in the 12th century and thenceforward, until about 1650, all printed calendars included astrological predictions. The study of the future through the stars was immensely popular. The field of astrology expanded into medicine: the human body, seen as a microcosm of the universe, seemed itself to be governed by the stars. Far left: each organ and member was influenced by a particular sign, as this 15th-century illustration indicates. Medicine relied on therapeutic almanacs, lunar calendars indicating the days favorable to cures such as bleeding, purges, and baths. "The moon in Aries is favorable for sowing, bathing, and taking remedies, especially for the head, throat, and chest, and for the phlegmatic," a Milanese almanac advised. Near left: a 17th-century Turkish manuscript illustrates the sign of Pisces.

became centers of economic activity. As commercial networks expanded, merchants needed to plan the delivery dates for products, and the schedules for the departure and arrival for ships. Bankers offered credit linked to the dates of the principal trade fairs, and bills of exchange became common. For the new businessmen time had a mercantile value; they had to quantify it in order to estimate prices. Greater rigor in references to time became necessary.

Some cities built great municipal planetary clocks that reproduced the motions of the heavens and earth, thus providing public astronomy lessons. Below: the working of a 15th-century clock is shown.

The first available instruments for tracking time were mechanical clocks, which appeared in Italy beginning in the late 13th century. Within the next century many of the cities of Europe had municipal clocks. Unlike church bells, which ring to announce matins, the mass, and vespers, clocks ring every hour. They "produce" time: quantitative time, composed of standard durations, objectively measurable. Civic clocks were useful for regulating the workday and were not subject to the control of the church. A secular, lay organization of time began to appear. Parallel to this, the calendar was coming into greater use. The more people were detached from natural and religious ways of measuring time, the more they relied on the calendar to track its duration.

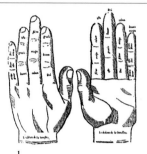

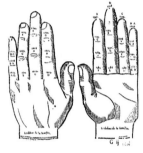

In the 15th and 16th centuries various mnemonic devices were used to explain the calendar. One method of finding a year's dominical letter proposed counting off the words of a formula on the fingers of the left hand. "People unable to count are no different from animals, but they should at least know what day it is," wrote the theologian Isidore of Seville in the 7th century. The hand calendar remained a useful means of learning dates until the early 17th century. Above: diagrams for a hand calendar from 1586.

The mastery of dates

The proliferation of mechanical clocks and the counting of the hours did not lead immediately to familiarity with the calendar. For most people, the rhythm of life still followed the sun—day and night, the seasons—and the liturgy—workdays, Sundays, and holy days. In villages the calendar for the coming week was announced from the pulpit each Sunday. It was a specialized calendar that did not measure the tempo of daily life, but organized church services and rituals. Neither dates nor the duration of the months were indicated. Days did not all have the same value; only Sundays and religious holidays mattered. These were designated by their names: for example, Quasimodo Sunday is the first Sunday after Easter (the name is a passage recited in the mass: *quasi modo geniti infantes…*); Rogation Days, also called Gang Days or Cross Week, were days of prayer and fasting on April 25 and the three days before the Feast of the Ascension; Maundy Thursday is the Thursday before Easter Sunday. Documents were dated in the same vague

"The clock is not merely a means of keeping track of the hours, but of synchronizing the actions of men…The clock, not the steam-engine, is the key-machine of the modern industrial age."
Lewis Mumford,
Technics and Civilization,
1934

way: St. John's Day, the first Sunday after St. Michael's, and so on.

Yet the increasing pace of trade in the West required the use of precise time references and it soon became essential to date commercial agreements by indicating day, month, and year.

This system of dating spread only slowly, since it was not easy to master. In the 14th century, only literate persons knew by heart the names of the months and their lengths. Students used mnemonic devices and counted on their fingers to recall the names of months and remember which feasts fell on which dates. These tricks were later taught in schools. In the space of two centuries precision was imposed; by the end of the 16th century, dates were common in written texts.

Almanacs spread awareness

The astrology vogue in turn led to a fad for almanacs. Physicians and astronomers alike tried their hands at forecasts and prognostications and wrote astrological

This 16th-century woodcut picture calendar, intended for the use of sailors, depicts the saints and their days with particular colors and images for each. The symbolism can be difficult to read, but we can spot the archangel Gabriel in March, St. George in April, St. James in May, and the lamb of St. John in June. Each day had a particular value and meaning.

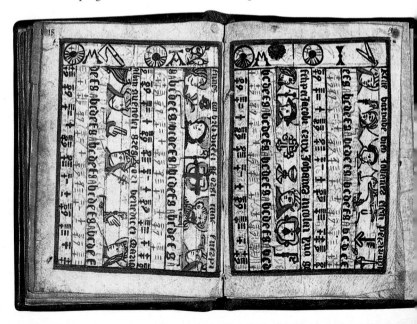

almanacs. An increasingly educated public, eager to know the future, acquired these works and became familiar with the calendar system through them. With the spread of printing in the 16th century, almanacs proliferated; they usually included a perpetual calendar together with an explanation of how time is subdivided into categories of season, month, day, and hour. The months were personified and compared to the different ages of life. This was useful in teaching dates.

Around 1550 a significant change occurred: almanac calendars began to present a quantified, numerical view of time. In more than half the copies in circulation, the days bore numbers. Where the year had once been simply a succession of feasts, it now presented a regular sequence of days, all clearly identified in an orderly and logical manner.

As almanacs circulated and people became familiar with dates, the ability to use the calendar to situate oneself in time was no longer the privilege of the church and chancery but became accessible to a growing population of educated laypeople. Individuals who were able to read could assert themselves and begin to seek to master time.

This was the Renaissance, the period in which the modern nation-states began to take shape. These new political entities had a strong motive to control the calendar. The Holy Roman Empire, in Germany, was the first to ordain that the entire kingdom start the year on January 1. Other rulers followed that initiative, in Spain, Denmark, the Low Countries, and France. Power over time began to change hands, from religious to secular authorities. And so the church responded.

Above: printed for the first time in 1491 and regularly republished until the 19th century, the *Shepherds' Kalendar* was a model for numerous other almanacs. It was an oddball compilation, meant for an urban, literate public, and blended a perpetual calendar with elements of astrology, medical advice, meteorological sayings, and articles of faith.

Attempted reforms

The Julian calendar had been a great improvement over previous systems, but its year was still 11 minutes, 14 seconds too long. This error, known since Ptolemy, posed problems in the calculation of the date of Easter, which is fixed on the first Sunday after the full moon that follows the vernal equinox. Accruing over several centuries, these extra minutes had shifted the date of the equinox from its true day toward winter. By the 16th century the difference amounted to ten days, with the equinox falling on March 11 instead of the official date of March 21. The calendar was doing a poor job of regulating time, for the social year no longer coincided with the solar year.

Most of the specialists responsible for determining the date of Easter in the Middle Ages were familiar with the shortcomings of the calendar. But they held the solid conviction that time came from God and the calendar reflected his will, so they were resigned to its flaws. Only a few learned men, such as the English philosopher Roger Bacon (c. 1220–92), argued for reform, and none of their attempts bore fruit. Jewish and Muslim scientists mocked the weaknesses of the Christian calendar.

By the early 16th century these imperfections were no longer tolerable. Demands for greater precision could no longer be ignored and the church took due notice. At the Fifth Lateran Council in 1514 Pope Leo X proposed a conference for a reform of the calendar. It received scant attention in most quarters, but—fatefully—caught the interest of one man, a Polish astronomer named Nicholas Copernicus (1473–1543). He set out to calculate the precise length of the year and in 1543, just before his death, published a treatise, *De Revolutionibus,* in which he expounded his so-called heliocentric theory, which held that the earth circled around the sun, rather than the sun around the earth. To support his case he provided a series of astronomical measurements that proved very accurate and were the foundation for a revision of the calendar.

The Franciscan monk Roger Bacon wrote to Pope Clement IV in 1267: "Our calendar is an offense against reason, an infamy for any healthy astronomy, the laughing-stock of all mathematicians." Below: the Sphere of Copernicus is an armillary sphere, a set of rings that reproduce the movements of the planets, or certain stars. The sun is placed at the center.

The Gregorian reform: ten days disappear

The church-sponsored Council of Trent (1545–63), reacting to the shocks of the Protestant Reformation, advocated an official reform of the calendar, but none was launched until 1575, when Pope Gregory XIII undertook the effort. The pope's primary objective was religious: he wanted to synchronize the church's time and natural time, stabilize the date of the equinox to keep Easter in the spring, and modify the lunar calendar so that Easter would not be celebrated with the new moon. Henceforth, a concern with accuracy prevailed over the force of tradition. God's time had to be precise, and that meant changing the Julian calendar.

In addition to this religious motive, the church also no doubt sought to regain its dominion over time. The period was marked by violent religious wars; at this critical moment the calendar was perceived as an essential measuring system that had slipped out of ecclesiastical control. The commission appointed to this task faced great structural difficulties, for the solar year does not have a complete number of days. To cope with this, they adopted a simple solution proposed by an Italian physician named Luigi Lillio (also called Aloysius Lilius, 1510–76): to delete three days from the calendar every 400 years by eliminating the bisextile (double-six) days from three out of four secular years. Only years divisible by 400

The German Jesuit Christophorus Clavius (1537–1612) was the driving force behind Pope Gregory's calendar reform and led the fight to enforce it. He was often faulted for his astronomical approximations and wrote five works to explain the reform and respond point by point to his critics. He became famous for his zeal, which was honored when the largest crater on the moon was named for him.

(that is, 1600, 2000, 2400) would keep 366 days. The other round numbers of centuries (1700, 1800, 1900, called secular numbers) would each have 365 days. To adjust for the accumulated extra days and make the shift to the new system, Lillio proposed cutting 10 days in a one-time correction. Finally, he suggested a new method for calculating the epact, the moon's age on January 1 each year.

The papal bull *Inter gravissimas,* signed on February 24, 1582, addressed all these points. Gregory XIII ordered that 10 days be eliminated from the calendar in October 1582, but without affecting the sequence of the days of the week, so that the day following Thursday, October 4 would be Friday, October 15. He set the beginning of the year for all of Christendom on January 1, the holy day of the circumcision of Christ.

This new calendar, known as the Gregorian, was sent to all Catholic states for immediate adoption. It was not a bolt from the blue; four years earlier, in 1578, Gregory XIII had sent a summary of Lillio's proposal to all heads of state and ecclesiastical dignitaries throughout Europe, to obtain their views.

The chosen solution was only approximate, but it had the great merit of being at once simple—modifying the Julian calendar only very slightly—and effective, since the Gregorian year was ahead of the exact year by only 26 seconds annually. (The exact year is the tropical year, measuring the earth's revolution around the sun from one vernal equinox to the next.) Currently the disparity is three hours; it will reach a full day sometime around the year 4700.

The Gregorian calendar may seem based on fragile premises. The scientific calculations were done with rudimentary instruments, the solution was approximate, and the reform itself was promulgated by a contested religious authority. And yet, four centuries later, it remains our calendar. It has even become the universal standard.

A long dispute

In Catholic Europe the Gregorian reform was quickly accepted by many countries, including Italy, Spain, and Portugal. Officially the change had occurred with the loss of 10 days in October 1582, but in France the transition

In 1576 Pope Gregory XIII assembled the members of his commission to reform the calendar. In this illustration one of them points out the error of the Julian calendar. The Council of Trent had already proclaimed the need to revise the calendar in 1563 and Gregory, elected pope in 1572, followed its directive. He considered the reform that bears his name one of the major accomplishments of his pontificate.

took place overnight from Sunday, December 9 to Monday, December 20, 1582. The change was adopted a few years later in Austria, Poland, Hungary, and in the Catholic states of the Holy Roman Empire (Germany).

But to Protestants the new calendar was by definition unacceptable. Some fifty years before, Martin Luther, learning of the existence of reform plans, had declared the calendar to be a matter for the civil authorities and not the church. Protestants considered the new calendar a "papist" measure, a sign of the pope's desire to master time and rule their lives. Such an act was all the more suspect in their eyes since it emanated from Gregory XIII, an ardent promoter of the Catholic Counter Reformation who had openly rejoiced at the slaughter of French Protestants in the St. Bartholomew's Day Massacre of

The Gregorian calendar reform was able to prevail over opposition from many quarters mainly because it coincided with the appearance of a new political and social order. It occurred during the Catholic Counter Reformation, which arose in response to the Protestant Reformation and the increasing power of nation-states. Intellectual life, too, was in ferment at the dawn of the scientific revolution of the 17th century.

August 24, 1572. "Protestants would rather disagree with the sun than agree with the pope," observed Kepler.

Thus, for political and religious reasons, the Protestant states of Europe remained faithful to the Julian calendar, which they called the Old Style (O.S.) calendar, to avoid any confusion with the New (N.S.) Gregorian. Britain, Sweden, and the American colonies maintained the division until the mid-18th century.

Indeed, England (and with it America) was one of the last countries in western Europe to adopt the Gregorian calendar. Resistance remained strong and it took all the energy, skill, and persuasive power of Lord Chesterfield, a former secretary of state, to persuade the English to join their neighbors. He worked for two years to mobilize the forces of public opinion, using the popular press and lobbying his colleagues. In 1752 the beginning of the year was moved from April 1 to January 1. Eleven days were eliminated in September so that the day after September 2 was September 14. This inspired cries of outrage and demands by the public to "give us back our eleven days!"

In this 1617 painting by Pieter Brueghel the Younger, depicting a tax collection, an almanac hangs on the wall behind the collector at far right. In the 17th century royal functionaries, legal officials, and merchants used almanacs to identify days by date. As professionals came to rely on calendars, a civil and secular means of measuring the year took root, as did the idea of the quantitative management of time.

The Orthodox Christian churches proved even more intractable than the Protestants. Not until the early 20th century did the states adopt the Gregorian calendar: Russia in 1918, Romania in 1919, Greece in 1924. Some churches offered partial compliance, but held to their own system for dating Easter. Today, a few Orthodox churches (in Russia, on Mount Athos in Greece, and in Jerusalem) keep the Julian calendar, which by now has fallen 13 days behind the Gregorian year.

Calendar time and clock time

A calendar enables us to orient ourselves in time, to identify the present with precision, mark off the past, and plan the future. Learning to use it can be quite a complex process. It took several centuries, in fact, for Westerners to go from simple dating in the 14th century to the sophisticated use of a personal datebook in the 18th.

By the close of the 16th century, educated people knew how to locate themselves in time by day, month, and year, with the help of calendars that numbered the days. Familiarity with dating changed common attitudes toward the idea of the past. As the philosopher Paul Ricoeur has noted, it is thanks to their registration in calendar time that memories built up by the collective consciousness become dated events. Without a calendar there can be no history in the modern sense.

Historical calendars began to appear in Germany in 1550. These were something like annals or chronicles published for an educated public. Calvinists and Catholics used these historical calendars in their vituperative quarrels, each citing various historical events to sustain their religious propaganda. Later, Jesuits and Jansenists also maintained their opposing polemical campaigns by citing calendars.

Until the 18th century the calendar was mainly to

For more than a hundred years Protestant Europe rejected the Gregorian reform, often fiercely. "We do not recognize this calendar maker as the shepherd of God's flock. He is just a howling wolf," wrote the German Protestant theologian Jacob Heerbrand in 1584 of Gregory XIII. Meanwhile the Protestants eliminated several of the new calendar's features, including the cult of the saints, fasting, and (at one time) Sunday church services, in the first step toward the secularization of time. Below: a 1752 English anti-reform print by William Hogarth proclaims, "Give Us Our Eleven Days."

This wall calendar from 1669 shows King Louis XIV granting an audience to foreign dignitaries and allegorical figures. The Sun King used the popularity of the calendar as propaganda, publishing it with pictures of himself in near-sacred guise and illustrated with important episodes of his reign, such as the births of heirs and military victories. In this ornate depiction of power every detail dramatizes the glory of the monarch: Louis appears at the top of the image to emphasize his superiority to other mortals. The ruling power controlled its own image as well as the printing of calendars, conflating the figure of royalty with that of time itself, and inventing a model for the illustrated calendar that was to have an illustrious future. Parisian printers specializing in this type of engraving were kept under close police surveillance and often subjected to police raids. Beginning in 1679 the Academy of Sciences, sponsored by the crown, published an annual official calendar that was reproduced in most almanacs.

structure the present and record the past. But during the Enlightenment it began to be used to organize the future. Around 1650 the first predictive calendars had appeared, with weekdays identified by date. This innovation, which enabled people to plan the future with precision, required about a century to become habitual and to appear in almanacs. The almanacs themselves continued to be popular and slowly evolved, as astrology and weather forecasts came to occupy less space in them. By the 18th century they were reaching a wider public.

In the 19th century datebooks, agendas, and strictly functional calendars began to appear with some regularity. These provide a visual diagram of future activities. The calendar had begun to take the form we know today, with days of the week and dates in the month in a list or grid arrangement, in the 17th century. Calendars were now updated every year and were widely used as literacy increased. Everyone who could read and write learned the divisions of time.

The evolution of the calendar was due to many factors, including progress in education and economic development. Clocks became common among the bourgeoisie in the 17th century and in the lower classes in the 18th. The watch, a miniature clock, made people more personally aware of time and helped bring them, in the words of the historian David Landes, from a state of time obedience (in which laborers or monks respond to a public bell) to one of time discipline (in which one orders one's life voluntarily, according to the movements of the clock).

New time for new people

The calendar has never lost its political significance. Various revolutionary movements have attempted to abolish the Gregorian calendar in order to make a clean break with a hated past and to reorder time according to a utopian vision.

A recent radical use of the calendar in an effort to erase the past and re-create the world occurred in Cambodia in 1975, when the dictator Pol Pot declared the Year Zero as part of his genocidal campaign to restructure every aspect of Cambodian life. The brutal experiment ended only with the downfall of his regime in 1979. But the Cambodian experiment has precedents.

Through the 17th century almanacs were restricted to a literate and educated audience, but by the 18th century they began to turn up in the countryside. They were sold by peddlers and soon became key elements in popular culture, especially the immensely popular astrology they contained. They were used in local schools to teach people to read. Most of them indicated the position of the moon and sun in the zodiac and included forecasts for health, climate, and various activities of work and pleasure. Above: a couple examines an astrological almanac in this engraving of 1771.

One of the most dramatic and interesting efforts to rethink the calendar occurred during the French Revolution. For twelve years, from 1793 to 1805, revolutionary France used its own homemade calendar. This ambitious reform had three objectives: to reject the preceding regime utterly, to provide secular holidays and a framework for life that would reflect the new society, and to rationalize all the systems of weights and measures, including the measurement of time.

The revolutionary French calendar was intended to demonstrate and institutionalize a historic turning point. It was meant to be irreversible, beginning anew with Year I. "We can no longer count the years when a king oppressed us as if they were a time we have lived through," wrote the playwright and revolutionatry politician Philippe-François Fabre d'Eglantine (1750–94), one of the forces behind the new calendar. His colleague Gilbert Romme (1750–95), kingpin in the calendar-reform commission, declared: "Time opens a new book in history…its new march [is] majestic and simple like equality!"

The new era in France began on September 22, 1792, with the downfall of the monarchy and the proclamation of the Republic. By a happy coincidence the 22d was the date of the autumn equinox. The revolutionists saw this as a good omen: civil equality echoed equality of night and day; thus history itself returned to nature.

The new calendar strove to be clear and exact, simple and universal. Like the reform of weights and measures, which imposed rational norms such as the gram and the meter, organized on the decimal system, it was part of a movement to ground public life in logic. The old system was deemed a monument to servitude and ignorance, riddled with anomalies such as unequal months and movable feasts. The new calendar used decimal computation, attuned to the motions of celestial bodies. All divisions of time smaller than the month were divisible by 10. The 12 months all had 30 days each and were further divided into periods of 10 days, called decades. There remained five intermediate days left over, which were placed at the end of the year, plus one extra day

"We made some watches…in which the day was broken down into decimal divisions. They can measure down to the 100,000th of a day, equivalent to the pulse-beat of a man of average height…moving at military double-time." Thus, in 1793, wrote Gilbert Romme, one of the architects of the French Revolutionary calendar, doing his best to link decimal divisions to the rhythms of nature. Objections to the 10-hour, 1,000-minute, 100,000-second day arose immediately. The revolutionists had underestimated popular resistance to the idea, as well as the difficulties of making the new timepieces. The act instituting the decimal time system was suspended on 18 Germinal, Year III (April 7, 1795). Above: an 18th-century watch.

every four years. A new organization of time it certainly was, but it essentially duplicated the system of ancient Egypt (which had 12 months of 30 days each, divided into 10-day periods, plus 5 extra days at year's end).

The revolutionary calendar was drawn up and promul-

The French Republican calendar established a new system of festivals. Above: the Feast of the Supreme Being (on 20 Prairial).

gated on October 5, 1793, during the period of the Revolution known as the Terror. The most extreme of the revolutionaries sought to purge French culture of what they saw as antiquated and corrupt influences, such as religion. Their calendar dechristianized time by suppressing Sunday, the Lord's day, and removing all the saints' days.

But what content was the new calendar to have instead? What symbolism was to be used? The revolutionary government chose a nature calendar. Each day was associated with a plant, animal, or agricultural instrument: tulip day, camomile day, and so on. December 25 became the day of the dog.

Months bore poetic, evocative names. Autumn included the months of Vendémiaire (wine harvest), Brumaire (foggy), and Frimaire (frosty); winter had Nivôse (snowy), Pluviôse (rainy), and Ventôse (blustery); spring had Germinal (seeding), Floréal (blossoming), and Prairial (open fields); summer had Messidor (harvest), Thermidor (heat), and Fructidor (fruitfulness).

Massive rejection

The revolutionary calendar was a resounding failure. It never succeeded in entering the common culture. Country folk in particular would not accept the disappearance of traditional feasts such as May Day, the bonfires of St. John's Day, or the days honoring their patron saints. Estranged from society and unable to penetrate the collective imagination, the new calendar disappeared in stages. In Year VIII the revolutionary holidays were suppressed. In Year X Napoleon Bonaparte restored Sunday as a day of rest, in order to heal the breach between the church and the revolutionary state. Finally, on 15 Fructidor, Year XIII (September 9, 1805), the calendar was officially abolished. Two reasons were given: it was not rational enough, and it was too nationalistic. This complete discrediting of a reform that had aspired to be scientific and universal had a great symbolic impact: the French Revolution was reintegrated into the process of history, as the philosopher Hannah Arendt noted. It had not contained enough meaning in its own right to usher in a new form of time. The Gregorian calendar was reinstated on January 1, 1806, a little over a year after Napoleon's coronation.

Fabre d'Eglantine wrote, "Our basic idea was to have the [new] calendar consecrate the agricultural system and to bring the nation back to this system by noting periods and fractions of years according to intelligible or visible signs taken from agriculture and rural economy." Together with the poet André Chénier and the painter Jacques-Louis David, Fabre invented the nomenclature of the Republican calendar. Engravings by Tresca were a direct application of his precepts, which are the earliest form of a very modern kind of illustrated calendar: the pin-up. Left: these pictures of the months Germinal, Ventôse, Messidor, and Brumaire were meant for a male audience, presenting the calendar and nature together as appealing, bountifully sexy female personifications. There were no overt messages, only suggestions. Overleaf: two French Victorian advertising calendars, one from a gas company, the other from a clothing shop. In the 19th century the calendar became a common publicity prop and was frequently given away to customers and subscribers, just as it is today. The material desires of a growing consumer society were linked with the dates and days of the year.

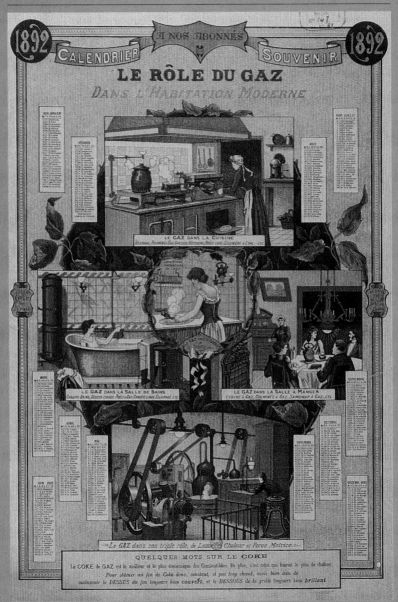

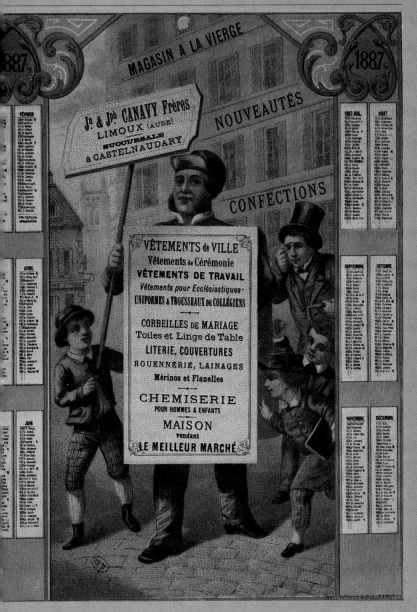

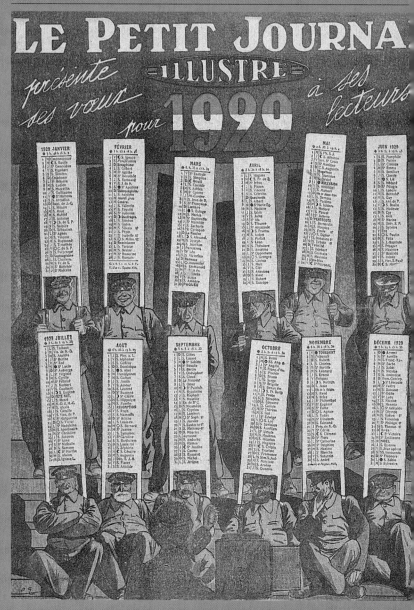

Januar 1952

S	M	D	M	D	F	S
		1	2	3	4	5
6	7	8	9	10	11	12
13	14	15	16	17	18	19
20	21	22	23	24	25	26

Far left: a calendar from *Le Petit journal illustré* (*The Little Illustrated Journal*), a popular French magazine. It was a free insert and bears the legend "With best wishes to our readers for 1929." Postmen who carry the months on their backs recall the role played by public institutions in promoting the notion of the calendar as a regulatory device. By choice or by force, in the factory or the army, workers and soldiers were inculcated with a strict sense of the discipline of time: arrive on time, work a full week, no missing the days after a holiday. Schoolchildren were taught in their history lessons to measure time in years and centuries, thus widening and organizing their temporal horizons. Near left: a 1952 German pin-up calendar, a 20th-century equivalent of the pictures on page 86. All kinds of images decorate yearly calendars, and the pin-up girl is only one of the most popular. Why is there such a profusion? Perhaps the printed calendar makes the future more concrete and less abstract. It is not only an aid to the memory, but also faithfully reflects the tastes of an era.

92

The calendar is subject to two contradictory forces. On the one hand, the measurement of time tends toward ever greater uniformity. Thus, the Gregorian calendar has gradually been adopted as a worldwide system, crossing among cultures. On the other hand, various political, social, and religious communities manipulate time to serve their needs and match their preferred rhythms. We live with many calendars, each a repository of collective memory.

CHAPTER 4
TOWARD A UNIVERSAL CALENDAR

"Time travels in divers paces with divers persons. I'll tell you who Time ambles withal, who Time trots withal, who Time gallops withal, and who he stands still withal."
William Shakespeare, *As You Like It*, Act III, scene 2, c. 1599–1600

Left: detail of a fresco by Enzo Cini, *Horse Hindquarters with Clock Face*, 1986–90; right: a souvenir desktop calendar with New York City statuettes.

Globalization leads to a single calendar

In the course of recent centuries the systems for recording time have gradually been reconciled. European colonial expansion from the 16th to the 19th century spread familiarity with Europe's calendar throughout the world. The European calendar was imposed on the colonized peoples of the Americas, Asia, and Africa; on gaining independence, they tended to retain it.

In the 20th century those countries in Eastern Europe that had maintained the Julian calendar all shifted to the Gregorian: Albania and Bulgaria in 1912, Russia in 1918, Romania and Yugoslavia in 1919, Turkey in 1926. The movement also reached the Far East, arriving in Japan in 1873, during the Meiji era, and in China, during the First Republic in 1912; this was later reconfirmed under communism by Mao Zedong in 1949. Each of these calendar changes accompanied political upheavals: the Russian Revolution, Westernization in Japan and Turkey, independence movements in Eastern Europe.

Today the Gregorian calendar is used nearly everywhere, although some states use a double calendrical system. In Israel Jewish and Gregorian calendars coexist. Newspapers carry both dates, as do official documents. Most of the Arab nations follow the same custom. Only a few of the Persian Gulf states maintain the Muslim calendar alone. A few specialized disciplines use their own dating systems. For example, paleontology refers to dates in the Common Era (AD or CE) as the present, but uses a

Above: the Hong Kong stock market in 2000. The coupling of computer technology and telecommunications unifies economic time. The financial system functions virtually non-stop around the world.

BP (before the present) dating system to accommodate extremely ancient dates.

Adopted, accepted, or tolerated by most nations, the Gregorian calendar has become the global time frame, the one used in international relations and economics. As international communication networks expand, growing more rapid and elaborate, they most often use the Western dating system, with only small local variations. (For example, in the United States the day is

Below: daily newspapers around the world follow a weekly rhythm. The 7-day week is now universal, although its composition has changed profoundly: it no longer consists of six profane days and one sacred, but of five workdays and two for leisure.

placed after the month in date citations, while in Britain it is placed before.)

The coordination of atomic and universal time

Along with the worldwide acceptance of the Gregorian calendar, time measurement itself has gradually been refined. As Western society developed, its need for a means of precisely synchronizing activities grew. In 1884 the hour underwent worldwide harmonization: the earth was sliced into 24 time zones by bands, called meridians, running north-south, parallel to the lines of longitude on the globe. The Greenwich meridian, at the Royal Observatory, near London, was chosen as the starting point for measurement: it is the Prime Meridian, longitude 0 0' 0". The International Date Line was established 180 degrees away from this meridian, exactly halfway around the world. On either side of this line, the hour is unchanged but the dates differ. Thus, 1 hour equals 15 degrees of latitude. To compare local time to the time at the Prime Meridian, one subtracts 1 hour from Greenwich time for every 15 degrees that one stands west of the Prime Meridian, or adds 1 hour for every 15 degrees that one stands east of it.

In 1911 Universal Time (UT) was instituted; this is Greenwich Mean Time (GMT), established at noon, to which 12 hours are added so that the day starts at midnight.

Precision was extended to the tiniest units of measure. In 1875 the International Bureau of Weights and Measures defined the second as 1/86,400th of the mean solar day. However, this definition has become obsolete. The earth's rotation was long considered a model of regularity, but the discovery of sizeable variations in the duration of the day has led to a search for more stable reference points. In 1956 the second became 1/31,556,925.9747 of the tropical year, with the year 1900 serving as a point of reference.

In 1967 atomic time made its appearance. The

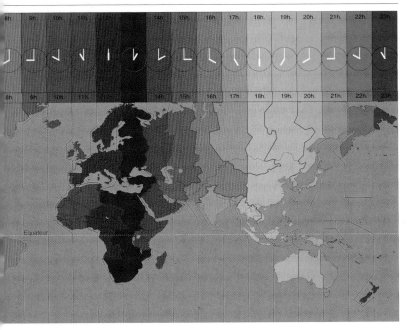

duration of the second was no longer tied to the earth's motion at all, but was measured according to the action of an atom of cesium. Thus, as the calendar historian David Ewing Duncan writes, "The year is no longer officially measured as 365.242199 days, but as 290,091,200,500,000,000 oscillations of Cs [cesium], give or take an oscillation or two." In this case the system of measurement is proving too perfect for the earth, whose rotation has a less regular duration. Coordinated Universal Time (UTC) is an updated name for GMT. It allows us to adjust atomic to solar time. Every six months the International Earth Rotation Service (IERS) can add or deduct one second from atomic time. Since 1972 the IERS has added a total of 22 seconds.

Slaves to the calendar

Today we distance ourselves more and more from nature's time. Central heating removes our sensitivity to cold weather, electric lighting frees us from night, and

Above: time zones around the world. The more we rely on long-distance communication systems, the more precise is the measurement of time. The heyday of the railroads in the 19th century led to the establishment of hourly time zones. The explosion in telecommunications led to atomic time. Far left: day, date, hour, minute, second, time zone, and phase of the moon: clock and calendar in one.

the urbanization that absorbs more than half of the world's population removes people from the rhythms of the natural world.

Time systems are also increasingly detached from nature, as the calendar grows more abstract. In many European countries the 52 weeks of the year are numbered consecutively, to permit clearer identification of them. The quarter, an abstract division, replaces the season, a visible one, in university and school programs and in assessments of the stock exchanges, national economies, and business. To measure longer time periods we have recourse to decades—the 1930s, the 1970s—and to periods of 100 years, centuries. According to the medievalist Jacques Le Goff these artificial divisions of time have tyrannized and distorted the study of history: "Everything has come to be couched in this artificial mold, as if the century were imbued with life, as if it had a unity, as if things had to change from one century to the next."

The complexity of life in an advanced society requires adept control of time. Because tasks follow one another relentlessly, a high premium is placed on punctuality and planning. Even leisure time fails to escape this framework. Vacations are arranged in advance and fall into ever shorter fragments, each with its agenda of activities. The sociologist Maurice Halbwachs has written, "Society, in obliging us to measure life by its standards,

"Today in our mechanized civilization we are dominated by the metered time of clocks. We submit to blackmail by the watch…Occasionally this makes us forget the time of duration. The clock thus becomes a kind of enemy that robs us of the dimension of memory."
Umberto Eco, interview in *Le Temps, vite* (*Fleeting Time*), exhibition catalogue, Paris, 2000

In the late 18th century the datebook, consisting of blank sheets of paper placed alongside the listing of the calendar dates, was just coming into use as a tool for organizing the future. At first people preferred to use datebooks like diaries, jotting down past events and memories in them. Today they have become an indispensable tool in public life, one that inserts individual time into collective time and helps us master the present. The electronic model, easier to update and more flexible, competes with the paper agenda. Our manner of depicting time is becoming virtual. Left: a written datebook; right: a photograph by André Kertesz of a bridge in Paris seen through a clock, 1929–32.

makes us more and more incapable of disposing of (and enjoying) our own life." Quantitative time invades qualitative time.

The illusion of the rational calendar

How good is the Gregorian calendar as an instrument of global coordination? As the demand for precision continues to grow, should we not develop a more efficient, rational calendar?

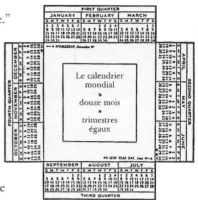

At regular intervals since its invention, critics have complained of the shortcomings of the Gregorian calendar. Divisions of the year are uneven, trimesters are not of uniform length, nor are months, and this causes political and economic problems. Worse yet, the days of the week shift year by year, so that a given date falls on the same day just once every 28 years. Movable holidays complicate planning. Finally, the Christian era provides an inexact means of counting years and an inaccurate millennium, since it has no year zero.

To remedy these drawbacks, plans for universal calendars have proliferated since the 19th century. In 1834 an Italian cleric named Marco Mastrofini proposed a calendar year of 364 days, always beginning on Sunday, January 1. The 365th day fell at the end, as an extra-calendrical holy day. Leap years had a second extra day. In 1849 the philosopher Auguste Comte proposed a "positivist calendar" based on this idea. It had thirteen months, each with 28 days and four weeks; the year ended with one or two "blank days" attached to no week. The proposal was burdened with ornamental names that made it cumbersome, and earned only brief attention.

Other plans divided the year into 4 quarters of 91 days each. Again, at year's end there were one or two days belonging to no week. Each quarter always began on a Sunday. What is striking here is that none of these proposals attacked the duration of the 7-day week. The only cycle that is totally artificial was the one no one

Attempts to create a universal calendar have always attracted a certain audience. Above: the calendar promoted by the World Calendar Association, whose director, Elisabeth Achelis, lobbied vigorously for it. In 1928 the Eastman Kodak Company adopted the calendar of the International Fixed Calendar League and organized its production and management in years of 13 months of 28 days each, a system it eventually abandoned only in 1990. But calendar reform has ceased to be the order of the day, as the trouble required to design and adopt such a change seems out of line with any expected benefits.

thought to modify, presumably because its religious and social significance had become thoroughly ingrained.

In the 20th century several groups turned once more to the idea of the universal calendar. The London-based International Fixed Calendar League, founded in 1923 by Moses B. Cotsworth, revived Comte's proposal for a 13-month year with 4 weeks per month and an extra Year Day at the end. The 13th month was to be called Sol. The IFCL had support from many American industrialists, especially George Eastman of the Eastman Kodak Company, as well as from the League of Nations, but the League was unable to achieve a consensus among member nations and in 1931 gave up the fight to adopt the concept. Meanwhile, in 1930 the World Calendar Association promoted a World Calendar Plan, but abandoned it in 1955.

On December 31, 1999 a Buddhist monk lights small candles in a Javanese temple hewn from lava to mark the start of the new millennium. He is thus celebrating the 3d millennium of the Christian deity by adapting a Chinese New Year's custom. This mixture of religious rites applied to a lay calendar is hardly surprising; all calendars assume something of a sacred quality.

Today the notion of a universal calendar seems more and more utopian, in particular because we now depend on computers whose programs are date-specific. Even so, reforms are still possible. The movable Christian feasts, Easter, Ascension, and Pentecost, could be fixed, and the Vatican has stated that it would not oppose such an idea. This proposal surfaces periodically, but without international political support it does not progress far.

Controversy over the Gregorian calendar has abated somewhat in our time. It has ceased to be seen as the instrument of Western technological colonialism or of an invasive and proselytic Christianity; in most contexts it is now accepted as a basically effective and neutral system for subdividing time.

The calendar, repository of collective memory

What becomes of other calendars when confronted with the dominance of the Gregorian? Are they condemned to extinction? On the contrary, many specialized calendars still survive. All cultures have their own periodicity, with

"The Jewish catechism is its calendar," wrote Rabbi Samson Raphael Hirsch in the late 19th century. Calendars are essential to organized religions because they register and repeat every year the rites that transmit the story of founding events. The Jewish calendar reenacts the formation of the Jewish people; the Christian liturgical year retraces the life of Christ. Above: a traditional Jewish prayer ritual with text, *talis* (prayer shawl), and *tefillin* (Torah boxes, worn on the arm, with straps wrapped around the hand).

important events inscribed in a localized calendar that constitutes a potent marker of identity. The calendar still unifies a group, tightens bonds among its members, and distinguishes it from others. Observance of specific calendar rituals connects individuals to one another. For example, the institution of the Shabbat, the pillar of Jewish religious life since the Babylonian Exile, allows Jews of the Diaspora to maintain a worldwide community. As the sociologist Eliatar Zerubanel has remarked, the Shabbat has helped the Jewish people to maintain a collective memory over many millennia and great opposing forces.

The calendar of the Coptic Christians of Egypt is one of the foundations of their cultural identity. In it is preserved the memory of the ancient Egyptians from whom they claim descent, and it preserves them from both the pressures of the majority Muslim culture within which they live and the alien forms of Christianity espoused by other churches. Christianized during the Roman period, the Copts separated from the official

The Coptic year begins with the month of Thoth, from the Greek name of the ancient Egyptian god who inscribed the names of the dead and protected scribes and mathematicians: the god of writing and knowledge, of computation and the calendar. The Alexandrian system (the Egyptian calendar as modified by Augustus Caesar) is also the basis of the calendars of the Parsees in India and the Ethiopians. The latter is very close to that of the Copts, but with its own month names. Its era begins on August 29, 7 BC. Above: a Christian holiday procession in Lalibela, Ethiopia.

church at the Council of Chalcedon in 451 by refusing to accept the dual nature of Christ as both human and divine. Later, during the Arab conquest, they resisted conversion to Islam. Their calendar, established in the 3d century AD, follows the ancient pharaonic Egyptian system: it has 12 months of 30 days each, which bear ancient names, plus 5 *epagomen* or intercalar days at the end of the year. The Copts adopted the adjustments of Julius Caesar, so that every four years they add an intercalar (Leap Year) day. Thus, they use a calendar that is both Egyptian and Julian. It also incorporates Christian elements: the first day of their Year 1 is August 29, 284, the year chosen in Alexandria in the reign of the Roman Emperor Diocletian as the point of departure for calculating the date of Easter. This calendar has not changed for 1,700 years; it is one of the essential elements of the Coptic minority's identity.

Because of this edict, Islam is one of the few religions whose feasts are not fixed at precise seasonal times. Left: Friday prayers in a mosque; above: *The Prophet Preaches Against the Intercalation of Months,* illumination in a 14th-century manuscript of the Arab astronomer al-Biruni (973–1048).

Muhammad chooses the moon

To maintain differences in a calendar, however small, reflects a desire to form an autonomous community. Shortly before his death in 632, the Muslim Prophet Muhammad introduced significant changes in the way time was counted among the Arabs. The traditional calendar of the Arab tribes was lunisolar. Muhammad banned the principle of intercalar months because, he

Muhammad named Friday as the holy day of the week, based on references in Genesis and the Jewish calendar and because on the sixth day God created humankind.

felt, such manipulations displeased God. He wanted to establish a calendar that would be impossible to change, one not linked to earth's rhythms. Only the moon could provide the necessary fixed points of reference.

In laying the groundwork for a lunar calendar Muhammad reinforced the identity of Muslim believers, isolated them from non-Muslim Arab tribes, and distinguished them clearly from the other religions of the Book: Judaism, with its lunisolar calendar, and Christianity, with its solar calendar. He accentuated this separation by selecting Friday as the day set aside for God. While Christians had chosen Sunday, the day after Shabbat, Muhammad chose the day before it.

This desire for dissociation was to have an unexpected consequence. Since the lunar calendar cannot be used in agriculture, Muslim farmers use the Christian system alongside their own. In the Maghreb (the Muslim countries of North Africa), rural calendars quite close to the Julian are composed of evocative sayings such as, "February sometimes smiles, sometimes drops sacks of snow," or, "Day and night will be equals when the fig leaf becomes as long as a mouse."

Survival of the ritual Chinese calendar

When the Gregorian calendar was grafted onto older cultures it did not always put down deep roots. In China two systems coexist, despite official support for the Western calendar. This remains abstract, for months have no names, but only numbers, and functional, since it is used mainly by

The traditional Chinese year, composed of twelve lunar months, is also divided for the convenience of farmers into 24 fixed periods, based on the solstices and equinoxes and bearing names from the agricultural calendar of northern China. The years are inscribed in 12-year cycles that have 6 wild animals (representing *yang,* the masculine, active, solar principle) and 6 domestic animals (representing *yin,* the feminine, passive, lunar principle). Numerous Asian peoples use this horoscope, which dates to the first centuries of the Common Era. Above: a Chinese astrological sign that is accompanied by the saying, "On the moon, the jade rabbit grinds medicinal plants in a mortar." In China and Japan the moon is a symbol of eternity. The hare with which it is often associated is making the elixir of eternity. Left: a 1996 Chinese calendar.

bureaucrats and businesses. The more popular calendar, the one that lends rhythm to collective life, is the traditional calendar, a system as sophisticated as the Jewish calendar. It is lunisolar, composed of twelve lunar months with supplemental months intercalated in obedience to the cycle of Meton. Since a calendar reform in 104 BC, the date of the New Year has not changed; it still falls on the day of the second new moon after the winter solstice. This solstice must always occur in the eleventh month. New Year's Day celebrations are by far the most important of the year; they last for nearly two weeks and mobilize the entire population. Other important festivals occur on doubled dates, such as the Feast of Dragons on the 5th day of the 5th month.

Despite great efforts, the Chinese authorities have not managed to efface the traditional calendar from people's minds, thus proving once again that a calendar cannot be decreed, but must compromise with tradition in order to endure.

Maintaining national spirit

Nations make their own calendars. They set administrative cycles: the budgetary year, fiscal year; they shape the political calendar: dates of elections, of sessions of Parliament or Congress or the courts, but their most important calendar function is to establish holidays, official festivals, and political celebrations. Holidays and vacations are never culturally neutral, for they indicate which aspects of the past have been chosen to remain in the collective

For the Chinese, February 5, 2000 marked the start of the Year of the Dragon. Chinese New Year is a spectacular celebration, halting business for several days. Below: New Year's lion dancers.

memory; they reflect a certain conception of the nation.

In Israel the chief national holidays underscore the state foundation myths. Holocaust Day, 27 Nisan, perpetuates the memory of the Shoah. It is followed one week later by Israeli Soldiers' Day and the day after that is Israel's national holiday. This cluster of three commemorations forms a meaningful sequence: from the defining tragedy of the Jewish people to the national combat that led to statehood. Here the calendar is clearly the instrument of a political symbolism.

Most nations celebrate holidays not linked to a religious calendar, commonly the date of their foundation or independence and a workers' holiday. Most also mark the anniversaries of important military victories or defeats that have defined their nationhood.

A national commemoration is both a ritual and an industry, mobilizing the past in the service of the present. Below: the 50th anniversary in June 1994 of the Normandy Invasion, the great Allied offensive of World War II, recalled a heroic moment of international unity and affirmed economic ties among the United States, France, Canada, Britain, and—perhaps most important—Germany, the former enemy.

Nepal, which uses Hindu solar and lunar calendars, celebrates National Unity Day (January 11), Martyrs' Memorial Day (January 29), National Democracy Day (February 19), Constitution Day (December 15), and the king's birthday (December 29). Barbados celebrates Errol Barrow Day (commemorating the leader of their independence, January 21), Heroes' Day (April 28), Emancipation Day (August 1), Kadooment Day (1st Monday in August), and Independence Day (November 30).

In European countries Remembrance Day or Armistice Day, on November 11, commemorates the dead of World War I and subsequent wars. The same day is called Veterans' Day in the United States. The United States has a plethora of federal (national) holidays: Martin Luther King, Jr. Day (3rd Monday in January),

Anniversaries, whether private, such as birth or marriage, or communal, such as memorial services, assume great importance in the social life of a culture. First introduced in the late 19th century, war commemorations are common today. The historian Pierre Nora writes: "The commemorative phenomenon was the concentrated expression of a national history, a rare and solemn moment…a transition from past to future. It has [now] become atomized. For each of the groups concerned it has become the thread running through the social fabric that will allow it, in the present, to establish a short cut to a past that is definitively dead…The commemoration has been freed from its traditionally assigned place; instead, the past event or moment as a whole has become commemorative…The past is no longer the guarantor of the future: that is the principal reason why memory has become the sole bearer of continuity. The solidarity of past and future has been replaced by the solidarity of the present and memory."

Pierre Nora,
Realms of Memory: The Construction of the French Past, vol. 3, *Symbols*, 1992

FÊTONS L'UNITÉ
1er MAI
1936
C.G.T.

LUTTONS LES 40 HEURES POUR LA PAIX
POUR Le Contrat Collectif POUR
Les Grands Travaux

Left: a 1936 French May Day poster, printed at the height of the international labor movement, declares, "Let us celebrate unity! Fight for a 40-hour work week, for collective bargaining, for public-works projects, for peace!" Every holiday has its rituals and its times of greater or lesser popularity. May Day, the international workers' day, and the American Labor Day are traditionally marked by parades with slightly militant overtones and festive public parties and picnics that owe something to older spring and harvest celebrations. At one time May 1 was a day of mass demonstrations, speeches, and political activism; in the old Soviet Union it was the great national holiday, with immense military displays. Today, as the history of the labor movement fades, such holidays have become ordinary days of leisure; in the same way, Sundays have lost some of their sacred character and are often seen primarily as days for rest.

Presidents' Day (3rd Monday in February), Memorial Day (last Monday in May), Flag Day (June 14), Independence Day (July 4), and Columbus Day (2nd Monday in October); however, not all states in the Union celebrate all of these, while every state also has its own local holidays. Both Canada and the United States celebrate a Thanksgiving Day, though on different dates (4th Thursday in
November in the United States and 2nd Monday in

Right: a jack-o-lantern recalls the lights, candles, and ghostly figures of other festivals near the winter solstice. Halloween reintroduces the awareness of the dead in the ambiguous form of a carnivalesque, transgressive holiday for children.

October in Canada).

Great Britain and the member countries of the British Commonwealth celebrate Commonwealth Day on the 2nd Monday in March, and three Bank Holidays, or long weekends. The first of these falls on May 1, or May Day, which is a civil holiday in many other Western countries as well. The date originated in the American trade-union movement. In the United States, labor agreements were dated starting May 1. For three years, from 1885 to 1888, American labor unions went on strike on that date in an effort to establish the right to an 8-hour workday. The idea spread internationally, and the date is now a national holiday in France, England, Russia, and elsewhere. Meanwhile, in the United States, the national workers' day was moved to the first Monday in September, and is called Labor Day. International Women's Day, March 8, had its roots in the early 20th-century labor movement, but has become a formal holiday in much of the world since it was formally recognized by the United Nations in 1975.

The calendar is not written in stone: new holidays appear from time to time. St. Valentine's Day and Halloween became popular in the 19th century, although their roots are older. Mother's Day is even more recent. Guy Fawkes Day in England, celebrated on November 5, dates to the Gunpowder Plot of 1605 and was originally political, but is now mainly a date for fireworks displays. Thanksgiving Day in the United States was inaugurated by George Washington in 1789.

Overleaf: the wheel of time, a 15th-century decoration on Gwalior Fort, India.

A consecration of collective life

It is not inappropriate to compare the calendar to the sedimentary strata of the earth's surface. Like the soil, the calendar forms a foundation of collective life. Like the soil, it comprises successive legacies, built up and worked upon by profound forces such as the movement to standardize time measurement, or the shift from religious society to secular society, both of which have modified it significantly. The calendar can be adapted by tenacious political rulers; it can be overturned or revised; it can incorporate new elements such as the need for exactitude, but the result will always depend on the local

DOCUMENTS

Traditional calendars

Although we do not have formal written calendars from the ancient world, people have always known ways to record, organize, and represent time.

Stellar guideposts

In antiquity, sailors and farmers planned their activities according to the stars. The early Greek writer Hesiod records advice for the planting, care, and harvesting of crops and names various stars, constellations, and gods as guides.

When Zeus has finished sixty wintry days after the solstice, then the star Arcturus leaves the holy stream of Ocean and first rises brilliant at dusk. After him the shrilly wailing daughter of Pandion, the swallow, appears to men when spring is just beginning. Before she comes, prune the vines, for it is best so.

But when the House-carrier [snail] climbs up the plants from the earth to escape the Pleiades, then it is no longer the season for digging vineyards, but to whet your sickles and rouse up your slaves. Avoid shady seats and sleeping until dawn in the harvest season, when the sun scorches the body. Then be busy and bring home your fruits, getting up early to make your livelihood sure…

But when Orion and Sirius are come into midheaven, and rosy-fingered Dawn sees Arcturus, then cut off all the grape-clusters…and bring them home. Show them to the sun ten days and ten nights: then cover them over for five, and on the sixth day draw off into vessels the gifts of joyful Dionysus [wine]. But when the Pleiades and Hyades and strong Orion begin to set, then remember to plough in season: and so the completed year will fitly pass beneath the earth.

Hesiod, *Works and Days,* ll. 564–77, 609–16, 8th century BC, trans. by Hugh G. Evelyn-White

Previous page: calendar in the form of a windmill, 19th century.

Understanding holidays

The classical Roman poet Ovid explains the religious calendar for the month of May.

The Scorpion will be visible from its middle in the sky, when we say that to-morrow the Nones will dawn.

When from that day the Evening Star shall thrice have shown his beauteous face, and thrice the vanquished stars shall have retreated before Phoebus, there will be celebrated an olden rite, the nocturnal Lemuria: it will bring offerings to the silent ghosts. The year was formerly shorter, and the pious rites of purification [*februa*] were unknown, and thou, two-headed Janus, wast not the leader of the months. Yet even then people brought gifts to the ashes of the dead, as their due, and the grandson paid his respects to the tomb of his buried grandsire. It was the month of May, so named after our forefathers [*maiores*], and it still retains part of the ancient custom. When midnight has come and lends silence to sleep, and dogs and all ye varied fowls are hushed, the worshipper who bears the olden rite in mind and fears the gods arises; no knots constrict his feet; and he makes a sign with his thumb in the middle of his closed fingers, lest in his silence an unsubstantial shade should meet him. And after washing his hands in clean spring water, he turns, and first he receives black beans and throws them away with face averted; but while he throws them, he says: "These I cast; with these beans I redeem me and mine." This he says nine times, without looking back: the shade is thought to gather the beans, and to follow unseen behind. Again he touches water, and clashes Temesan bronze, and asks the shade to go out of his house. When he has said nine times, "Ghosts of my fathers, go forth!" he looks back, and thinks that he has duly performed the sacred rites…

The times are unsuitable for the marriage both of a widow and a maid: she who marries then, will not live

Medallions depicting the months of April and December, 17th century.

long. For the same reason, if you give weight to proverbs, common folk say 'tis ill to wed in May. But these three festivals fall about the same time, though not on three consecutive days.

Ovid, *Fasti* (*The Holidays*),
c. AD 4,
book 5, ll. 417–44, 489–92
trans. Sir James G. Frazer

Dating in the 16th century

The historian Lucien Febvre notes that in the 16th century dating was unusual: lived time took precedence over measured time.

[Most people] don't even know their age exactly—that is, not counting the historic personages of the time, who leave us a choice among three or four recorded dates of birth, sometimes several years apart. When was Erasmus born? He did not know; he only knew that the event had taken place on the eve of St. Simon's and St. Luke's Day.

In what year was Jacques Lefebre d'Etaples born? Efforts have been made to deduce the date based on only the vaguest references. What about François Rabelais's birth date? He didn't know. And Martin Luther? We're not sure. One tended to know the month— but the year itself was very badly organized, since the vernal equinox had gradually slid from March 21 back to March 11. The family remembered the little one came into the world at the time of the harvest of wheat or grapes; or there was snow; or maybe it was the month when corn ripened…"When

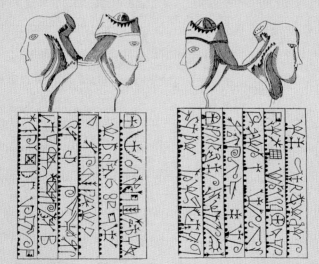

In this calendar discovered in 1732 in the Breton chateau of Coëdic, saints and liturgical holidays are represented by symbols. For example, in the second column from the right, near the top, is a square grill. This is the grill on which St. Lawrence was martyred, and represents August 10, St. Lawrence's Day. The calendar probably dates to the end of the 17th century; its authorship is unknown.

the wheat has already begun to fall,…when the stems begin to poke up"; John Calvin gives us this date, bucolically precise, in its way. Then family tradition became frozen [by the introduction of dating]: Francis was born on November 27 and John on January 12—how cold it was when he was taken to the baptismal font!

Lucien Febvre,
*Le Problème de l'incroyance
au XVIe siècle:
La religion de Rabelais*
(*The Problem of Disbelief
in the 16th Century:
The Religion of Rabelais*), 1968

A horticultural calendar committed to memory

Even in the modern era oral calendars like Hesiod's are still sometimes used. This one, recorded in the 1970s in eastern France, describes traditional rules for planting that depend on the moon and the saints.

The March moon—starting at the last new moon of the month—is cold and bad-tempered, like the dry, freezing winds that come with it. Beware of it, and of the red moon a month later. Once you've passed this first, malevolent moon, begin the spring sowing, which stops only for the red moon of April, last new moon of the month. These red moons are said to be sterile…On the other hand, the August moon associated with St. Lawrence Day is benevolent and favorable to planting. Under this kindly moon you can begin the autumn planting, hurrying to cut the weeds and thorns, which won't grow back.

Aside from these strange moons, disastrous or favorable, the phases of stars throughout the seasons also influence farming.

Depending on the type of plant, we sow, or cut again during the new moon, that is, when it is rising, or by the old moon or a waning moon. The first favors growth in height and the propagation of everything growing above the ground, while the second maintains underground life and fertilizes creeping plants. "In the new moon everything seeds, everything flowers; in the old moon everything yields. One may plant everything that climbs, all the little seeds at the rising moon; peas put out more pods, as do beans…Spinach, carrots, potatoes must be planted by the old moon or they will flower all year long; lettuce and cabbages will form a head, onions will harden…If you plant them in the new moon, they shoot straight up and don't form bulbs."

The phases of the moon organize cultivated plants into two major categories: those that grow up into the air and those that lie low to the ground or creep. The saints of the religious calendar protect each plant specifically: "On St. Agatha's Day, February 5, we plant spring lettuce; at Mardi Gras plant parsley; on May 3, the day of the Holy Cross, plant beans, and also on Rogation Days: they grow in profusion, but on May 31, St. Petronilla's Day, one bean will yield a thousand; on St. Medard's Day, a billion; but on St. Barnaby's Day you'll only get a thousand…"

F. Zonabend,
*La Mémoire longue:
Temps et histoire au village*
(*Long Memory: Time and
History in a Village*),
1980

L e *Kalendrier des Bergers*, known in English as *The Shepherds' Kalendar*, was first printed in 1491 and republished 40 times over the next 300 years.

Traditionally, the months were depicted as figures carrying out set tasks linked to the works and days of the rural world.

Views of the Gregorian reform

The Gregorian calendar was accepted by various nations at widely differing times. Naturally, writers commented upon it.

In the American colonies, 1752

The American inventor, wit, and scientist Benjamin Franklin (1706–1790) began writing and publishing an annual pamphlet called Poor Richard's Almanack *in 1733, when he was 27. The booklet was a calendrical mixture of weather forecasts for the year, good advice, proverbs, poems, and maxims. A version of it is still published annually in the United States today. The title page of the first issue reads:*

Poor Richard, 1733, an Almanack for the Year of Christ 1733, being the first after Leap Year:

And makes since the Creation	Years
By the Account of the Eastern Greeks	7241
By the Latin Church	6932
By the Computation of W.W.	5742
By the Roman Chronology	5682
By the Jewish Rabbies	5494

Wherein is contained

The Lunations, Eclipses, Judgment of the weather, Spring Tides, Planets, Motions & mutual Aspects, Sun and Moon's Rising and Setting, Length of Days, Time of High Water, Fairs, Courts, and observable Days. Fitted to the Latitude of Forty Degrees, and a Meridian of Five Hours West from London, but may without sensible Error, serve all the adjacent Places, even from Newfoundland to South-Carolina.

The Almanack *for 1752 contained a commentary on the history of the calendar and the then-recent (and quite controversial) decision of the British authorities in England to adopt the Gregorian reform.*

KIND READER,

Since the King and Parliament have thought fit to alter our Year, by taking eleven Days out of September, 1752, and directing us to begin our Account for the future on the First of January, some Account of the Changes the Year hath heretofore undergone, and the Reasons of them, may a little gratify thy Curiosity.

The Vicissitude of Seasons seems to have given Occasion to the first Institution of the Year. Man naturally curious to know the Cause of that Diversity, soon found it was the Nearness and Distance of the Sun; and upon this, gave the Name Year to the Space of Time wherein that Luminary, performing his whole Course, returned to the same Point of his Orbit.

And hence, as it was on Account of the Seasons, in a great Measure, that the Year was instituted, their chief Regard and Attention was, that the same Parts of the Year should always correspond to the same Seasons; i.e. that the Beginning of the Year should always be when the Sun was in the same Point of his Orbit; and that they should keep Pace, come round, and end together.

This, different Nations aimed to attain by different Ways; making the Year to commence from different Points of the Zodiac; and even the Time of his Progress different. So that some of their Years were much more perfect than others, but none of them quite just; i.e. none of them but whose Parts shifted with Regard to the Parts of the Sun's Course.

It was the Egyptians, if we may credit Herodotus, that first formed the Year, making it to contain 360 Days, which they subdivided into twelve Months, of thirty Days each.

Mercury Trismegistus added five Days more to the Account.—And on this Footing Thales is said to have instituted the Year among the Greeks. Tho' that Form of the Year did not hold throughout all Greece. Add that the Jewish, Syrian, Roman, Persian, Ethiopic, Arabic, &c. Years, are all different.

In effect, considering the poor State of Astronomy in those Ages, it is no Wonder different People should disagree in the Calculus of the Sun's Course. We are even assured by Diodorus Siculus, Plutarch, and Pliny, that the Egyptian Year itself was at first very different from what it became afterwards.

According to our Account, the Solar Year, or the Interval of Time in which the Sun finishes his Course thro' the Zodiac, and returns to the same Point thereof from which he had departed, is 365 Days, 5 Hours, 49 Minutes; tho' some Astronomers make it a few Seconds, and some a whole Minute less; as Kepler, for Instance, who makes it 365 Days, 5 Hours, 48 Minutes, 57 Seconds, 39 Thirds. Ricciolus, 365 Days, 5 Hours, 48 Minutes. Tycho Brahe, 365 Days, 5 Hours, 48 Minutes.

The Civil Year is that Form of the Year which each Nation has contrived to compute Time by; or the Civil is the Tropical Year, considered as only consisting of a certain Number of whole Days; the odd Hours and Minutes being set aside, to render the Computation of Time in the common Occasions of Life more easy.

Hence as the Tropical Year is 365 Days, 5 Hours, 49 Minutes; the Civil Year is 365 Days. And hence also, as it is necessary to keep Pace with the Heavens, it is required that every fourth Year consist of 366 Days, which would for ever keep the Year exactly right, if the odd Hours of each Year were precisely 6.

The ancient Roman Year, as first settled by Romulus, consisted of ten Months only; viz. I. March, containing 31 Days. II. April, 30. III. May, 31. IV. June 30. V. Quintilis, 31. VI. Sextilis, 30. VII. September, 30. VIII. October, 31. IX. November, 30. X. December, 30; in all 304 Days; which came short of the Solar Year, by 61 Days.

Hence the Beginning of Romulus's Year was vague, and unfixed to any precise Season; which Inconvenience to remove, that Prince ordered so many Days to be added yearly, as would make the State of the Heavens correspond to the first Month, without incorporating these additional Days, or calling them by the Name of any Month.

Numa Pompilius corrected this irregular Constitution of the Year, and composed two new Months, January and February, of the Days that were used to be added to the former Year. Thus, Numa's Year consisted of twelve Months; viz. I. January, containing 29 Days. II. February, 28. III. March, 31. IV. April, 29. V. May, 31. VI. June, 29. VII. Quintilis, 31. VIII. Sextilis, 29. IX. September, 29. X. October, 31. XI. November, 29. XII. December, 29; in all 355 Days; which came short of the common Solar Year by ten Days; so that its Beginning was vague and unfixed.

Numa, however, desiring to have it fixed to the Winter Solstice, ordered 22 Days to be intercalated in February every second Year, 23 every fourth, 22 every sixth, and 23 every eighth Year. But this Rule failing to keep Matters

even, Recourse was had to a new Way of Intercalating; and instead of 23 Days every eighth Year, only 15 were added; and the Care of the whole committed to the Pontifex Maximus, or High Priest; who, neglecting the Trust, let Things run to the utmost Confusion. And thus the Roman Year stood till Julius Caesar made a Reformation.

The Julian Year, is a Solar Year, containing commonly 365 Days; tho' every fourth Year, called Bissextile, contains 366.—The Names and Order of the Months of the Julian Year, and the Number of Days in each, are well known to us, having been long in Use.

The astronomical Quantity, therefore, of the Julian Year, is 365 Days, 6 Hours, which exceeds the true Solar Year by 11 Minutes; which Excess in 131 Years amounts to a whole Day.—And thus the Roman Year stood, till the Reformation made therein by Pope Gregory.

Julius Caesar, in the Contrivance of his Form of the Year, was assisted by Sosigenes, a famous Mathematician, called over from Egypt for this very Purpose; who, to supply the Defect of 67 Days which had been lost thro' the Fault of the High Priests, and to fix the Beginning of the Year to the Winter Solstice, made that Year to consist of 15 Months, or 445 Days; which for that Reason is used to be called *Annus Confusionis,* the Year of Confusion.

This Form of the Year was used by all Christian Nations, till the Middle of the 16th Century; and still continues to be so by several Nations; among the Rest, by the Swedes, Danes,&c. and by the English till the second of September next, when they are to assume the Use of the Gregorian Year.

The GREGORIAN YEAR is the Julian Year corrected by this Rule; that whereas

The biblical book of the Revelation of St. John the Divine describes the end of time and is the basis for many millenarian ideas. Above: a medieval manuscript illumination depicts the end of days.

on the common Footing, every Secular or Hundredth Year, is Bissextile; on the new Footing, three of them are common Years, and only the fourth Bissextile.

The Error of eleven Minutes in the Julian Year, little as it was, yet, by being repeated over and over, at length became considerable; and from the

Medallions depicting the months of September and November, 17th century.

Time when Caesar made his Correction, was grown into 13 Days, by which the Equinoxes were greatly disturbed. To remedy this Irregularity, which was still a growing, Pope Gregory the XIII called together the chief Astronomers of his Time, and concerted this Correction; and to restore the Equinoxes to their Place threw out the ten Days that had been got from the Council of Nice, and which had shifted the fifth of October to the 15th.

In the Year 1700, the Error of ten Days was grown to eleven; upon which the Protestant States of Germany, to prevent further Confusion, accepted the Gregorian Correction. And now in 1752, the English follow their Example.

Yet is the Gregorian Year far from being perfect, for we have shewn, that, in four Centuries, the Julian Year gains three Days, one Hour, twenty Minutes: But it is only the three Days are kept out in the Gregorian Year; so that here is still an Excess of one Hour, twenty Minutes, in four Centuries; which in 72 Centuries will amount to a whole Day.

As to the Commencement of the Year, the legal Year in England used to begin on the Day of the Annunciation; i.e. on the 25th of March; tho' the historical Year began on the Day of the Circumcision; i.e. the first of January, on which Day the Italian and German Year also begins; and on which Day ours is to begin from this Time forward, the first Day of January being now by Act of Parliament declared the first Day of the Year 1752.

At the Yearly Meeting of the People called Quakers, held in London, since the Passing of this Act, it was agreed to recommend to their Friends a Conformity thereto, both in omitting the eleven Days of September thereby directed to be omitted, and beginning the Year hereafter on the first Day of the Month called January , which is henceforth to be by them called and written, The First Month, and the rest likewise in their Order, so that September will now be the Ninth Month, December the Twelfth. This Act of Parliament, as it contains many

Matters of Importance, and extends expressly to all the British Colonies, I shall for the Satisfaction of the Publick, give at full length: Wishing withal, according to ancient Custom, that this New Year (which is indeed a New Year, such an one as we never saw before, and shall never see again) may be a happy Year to all my kind Readers.

Richard Saunders
[pseudonym for Benjamin Franklin],
Poor Richard's Almanack, 1752

In England, 1646

The 17th-century English physician and philosopher Sir Thomas Browne was fond of correcting superstitions and popular misconceptions about the natural world and drawing moral conclusions from them. He wrote at a time when much of Europe had adopted the Gregorian calendar, while Britain still kept to the Julian.

OF SOME COMPUTATIONS OF DAYES AND DIDUCTIONS OF ONE PART OF THE YEAR UNTO ANOTHER.

There are certaine vulgar opinions concerning dayes of the yeare and conclusions popularly deduced from certaine dayes of the month; men commonly beleeving the dayes encrease and decrease equally in the whole yeare, which notwithstanding is very repugnant unto truth; For they encrease in the month of March, almost as much as in the two months of January and February; and decrease as much in September, as they doe in July and August: For the dayes encrease or decrease according to the declination of the Sun; that is, its deviation Northward or Southward from the Æquator. Now this digression is not equall, but neare the Æquinoxiall

intersections, it is right and greater, neare the Solstices more oblique and lesser. So from the eleventh of March the vernall Æquinox unto the eleventh of Aprill the Sun declineth to the North twelve degrees; from the eleventh of Aprill unto the eleventh of May but eight, from thence unto the fifteenth of June, or the Summer Solstice but three and a half; all which make twenty three degrees and an halfe, the greatest declination of the Sun.

And this inequality in the declination of the Sun in the Zodiacke or line of life, is correspondent unto the growth or declination of man; for setting out from infancie we encrease not equally, or regularly attaine to our state or perfection; nor when we descend from our state, is our declination equall, or carryeth us with even paces unto the grave. For, as Hippocrates affirmeth, a man is hottest in the first day of his life, and coldest in the last; his naturall heate setteth forth most vigorously at first, and declineth most sensibly at last. And so though the growth of man end not

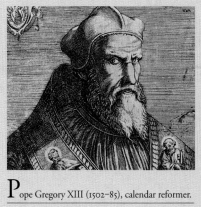

Pope Gregory XIII (1502–85), calendar reformer.

perhaps until twenty one, yet is his stature more advanced in the first septenary then in the second, and in the second, more then in the third, and more in the first seven yeares, then in the fourteene succeeding; for, what stature we attaine unto at seven yeares, we do sometimes but double, most times come short of at one and twenty. And so do we decline againe; for in the latter age upon the Tropick and first descension from our solstice, we are scarce sensible of declination; but declining further, our decrement accelerates, we set apace, and in our last dayes precipitate into our graves…

There are also certaine popular prognosticks drawne from festivals in the Calendar, and conceaved opinions of certaine dayes in months, so is there a generall tradition in most parts of Europe, that inferreth the coldnesse of succeeding winter from the shining of the Sun upon Candlemas day…

So is it usual amongst us to qualifie and conditionate the twelve months of the yeare, answerably unto the temper of the twelve dayes of Christmasse, and to ascribe unto March certaine borrowed dayes from Aprill; all which men seeme to beleeve upon annuall experience of their own, and the receaved traditions of their forefathers.

Now it is manifest, and most men likewise know, that the Calendars of these computers, and the accounts of these dayes are very different; the Greekes dissenting from the Latins, and the Latins from each other; the one observing the Julian or ancient account, as great Britaine and part of Germany; the other adhering to the Gregorian or new account, as Italy, France, Spaine, and the united Provinces of the Netherlands. Now this later account

by ten dayes at least anticipateth the other; so that before the one beginneth the account, the other is past it; yet in the severall calculations, the same events seeme true, and men with equall opinion of verity, expect and confesse a confirmation from them all…

And thus may [those] easily be mistaken who superstitiously observe certaine times, or set downe unto themselves an observation of unfortunate months, or dayes, or howres; As did the Ægyptians, two in every month, and the Romans, the dayes after the Nones, Ides, and Calends. And thus the Rules of Navigators must often faile, setting downe, as Rhodiginus observeth, suspected and ominous dayes, in every month, as the first and seventh of March, the fift and sixt of Aprill, the sixt, the twelfth and fifteenth of February. For the accounts hereof in these months are very different in our dayes, and were different with severall nations in Ages past; and how strictly soever the account be made, and even by the selfe same Calendar, yet is it possible that Navigators may be out. For so were the Hollanders, who passing Westward through *fretum le Mayre,* and compassing the Globe, upon their returne into their Owne Countrey, found that they had lost a day. For if two men at the same time travell from the same place, the one Eastward, the other Westward round about the earth, and meet in the same place from whence they first set forth; it will so fall out, that he which hath moved Eastward against the diurnall motion of the Sun, by anticipating daylie something of its circle with his owne motion, will gaine one day;

but he that travelleth Westward, with the motion of the Sun, by seconding its revolution, shall lose or come short a day...

Sir Thomas Browne,
Pseudodoxia Epidemica (Vulgar Errors),
VI: iv, 1646

In France, 1580

The French Renaissance philosopher Michel de Montaigne saw no difficulty with the change to the Gregorian calendar when it occurred in his country.

Two or three years ago they shortened the year by ten days in France. How many changes followed this reform! It was like moving heaven and earth at the same time. Nevertheless, nothing really moved from its place: my neighbors find the hour of their sowing, of their harvest, the opportunity for business dealings, the unfavorable and beneficial days, all at the very place where they'd always been. We neither felt an error in our customs, nor did we feel the correction. There is so much uncertainty all about, so much that is perceived indistinctly, dark and confused. It is said that the change could have been carried out in a less disturbing way: one could follow Augustus's example and for a few years leave off the 366th [Leap Year] day, which is a day of problems and turmoil, until the debt of time was settled exactly. (For that was not even accomplished by the present correction, so that we still remain a few days in arrears.) And if, by the same token, we could control the future, ordering that through a given number of years this extra day would still be deleted, then the error would be no more than 24 hours. We have no other way of

New Year's Eve in Times Square, New York, 2000.

counting time than years. The world has been using them for so many centuries!

Michel de Montaigne (1533–92),
Essays, vol. 3, 1580

Rethinking the calendar

It is no secret that the calendar has political uses and can be a tool of both governments and their critics.

Saturn will be retrograde…

François Rabelais (c. 1494–1553) wrote a delightful satire on almanacs and their predictions.

This year will bring so many eclipses of the Sun and Moon that I am afraid (and rightly too) that the result will be starvation for our purses and perturbation for our senses.

Saturn will be retrograde, Venus direct. Mercury inconsistent. And a whole pile of other planets will refuse to do your bidding.

Thus for this year the pigs will run sideways and rope makers backward; ladders will climb on benches, spindles on ladders, and bonnets on hats. Testicles will expand for many; through lack of a game bag, fleas will be black, for the most part. Lard will flee from peas during Lent; the belly will precede us, the backside will be first to sit down; you won't be able to find the prize in the Three Kings' Cake, nor to find an ace in a flood; dice won't respond as desired despite all manner of flattery, and the luck we request won't appear often, while beasts will speak in various

M edallions depicting the months of June and February, 17th century.

places. The Lent Observer will win his trial; one part of the world will disguise itself to fool the other, and will run through the streets like idiots, out of their senses. Never was such disorder seen in nature. And this year will see the formation of more than XXVII irregular verbs…If God does not help us, we'll have rather little business, but on the other hand, if He is for us, nothing can harm us…

OF THE MALADIES OF THIS YEAR

This year the blind will see very little, the deaf will have trouble hearing, the mute will barely speak: the wealthy will do a bit better than the poor, and the healthy better than the sick. Several Sheep, Bulls, Pigs, Birds, Chickens, and Ducks will die but there will be less cruel mortality among the Monkeys and Dromedaries. Old age will be incurable this year, because of prior years. Those with pleurisy will have bad pain in the side, while those who have stomach trouble will go often to the chair with a hole in the seat; catarrhs will descend this year from the head to the lower members: eye trouble will be most unfavorable to sight…and almost universally there will be a quite horrible, frightful malady: malicious, perverse, frightful, and unpleasing, which will leave people quite astonished, many not knowing which way to turn, and often they will descend into reverie, arguing eloquently about the Philosopher's stone and the ears of Midas. I tremble with fear thinking of it, for I tell you that it will be epidemic…

Last year's comet is awaited, along with Saturn's retrograde motion. A great clown, catarrhous and encrusted, will die in hospital. Death of the same

will cause horrible sedition between cats and rats, between dogs and hares, between falcons and ducks, monks and eggs.

François Rabelais,
*Pantagrueline Prognostication
pour l'an 1533*
(*Pantagruelean Prognostication
for the Year 1533*)

Who is not stirred by those lovely names?

There are arguments in favor of an altered calendar. The 19th-century historian Jules Michelet wrote an impassioned tribute to the experiment in calendar reform attempted by the French Revolution.

On September 20th, two days before the anniversary of the founding of the Republic, [Gilbert] Romme read to the Convention the draft of the Republican Calendar, which was adopted on October 5th. For the first time in history, man had a true measure of time…

Romme's calendar is full of his stoical genius and austere faith in pure Reason. Nothing that could give rise to idolatry: neither saints nor heroes. For the month, he had universals, Justice, Equality, etc. Just two of the months embodied sublime moments [relating to the history of the Revolution]: June, the *Serment du Jeu de Paume* [the Tennis Court Oath] and July, *Bastille* [the prison whose fall signaled the start of the Revolution].

The rest were names or numbers. Days and decades by numbers. Day followed day, equal in duty, equal in work. Time took on the unvarying guise of eternity.

It was an austere calendar, and yet well received, for people hungered and thirsted after truth. On October 10th, in Arras, the northern departments held a prodigious festival, astronomical and mathematical; earth imitated the heavens. No less than twenty thousand took part, all figuring in a gigantic pageant of the movement of the year. And it all took place so close to the enemy, in solemn expectation of the six days which preceded the battle which liberated France!…Over there was idolatrous Belgium and the army which sought to restore false gods; here was France, pure, powerful and peaceable, acting out the sacred pageant of Time, celebrating the new era, the greatest the world had ever known.

The twenty thousand were divided by age into twelve groups, each representing a month. There was a procession of the entire year, figured in all its variety by the infinite variety of human faces: young and laughing; then ripe, grave; and finally seeking their merited rest. Those who had conquered life and passed their eightieth year were formed into a sacred little group of their own, symbolizing the complementary days which completed the Republican calendar. The extra day added every fourth year was represented by a man who had reached his hundredth year; he walked under a canopy. Behind the old men bent over their staffs walked all the little children; the new year following the old, as new generations follow those who go to the grave…

Could this austere calendar and these infinitely pure festivals, all reason and heart and cold comfort for the imagination, replace the gibberish of the old baroque Almanac, gaudy with the hundred colors of idolatry, freighted with legendary feast days, bizarre names men mumbled without understanding…? The Convention felt the popular soul deserved less abstract nourishment; it kept Romme's scientific base but

changed the nomenclature. The ingenious Fabre d'Eglantine, in a pleasant little book written in more peaceful times, in 1783 [*Natural History in the Course of the Seasons*], had suggested a true calendar in which nature itself, in the language of its fruits and flowers, the beneficent unfolding of its motherly gifts, provided the periods of the year. The days were named after harvests, so that the whole calendar is a sort of manual for the farmer; day by day his life is linked with his toil in the fields. What could be more suitable for the agricultural nation France then was? The months, named after the cycle of weather and harvests, were so felicitous and expressive, so charming, that all took them straight to their hearts, where they are still enshrined. Today, they remain part of our heritage, one of those ever-living creations in which the Revolution survives and ever will. Who is not stirred by those lovely names? If the unhappy Fabre saw no more than four months of his calendar; if, arrested in Pluviôse, he died with Danton in Germinal, his death, too cruelly avenged in Thermidor, could not deprive him of his immortality, for he it was who understood nature and sang the song of the year.

The implications of these changes were immense. At stake was nothing less than a change of religion.

The almanac is a more serious business than empty-headed people believe. The struggle between the two calendars, the republican and the catholic, is the struggle of *tradition,* of the *past,* against the eternal *present* of nature and calculation.

Nothing irked yesterday's men more. "What use is this calendar?" Bishop

Grégoire one day asked Romme. Romme answered, "At least it rids us of Sundays."

Jules Michelet (1798–1874), *History of the French Revolution,* vol. 7, book 14, ch. 2, 1847, trans. by Keith Botsford

The Soviet week

In 1918, during the Bolshevik Revolution, the Julian calendar was abolished in the new Soviet Union, replaced by the Gregorian. In 1929 came a new modification, inspired by the old French Revolutionary calendar: the year was divided into 12 months of 30 days each, to which were added 5 or 6 extra days. Each month had 6 weeks of 5 days. The days kept their names, except Saturday and Sunday, which disappeared. The reform served a dual objective: to optimize industrial output by instituting a continuous production cycle, and to dechristianize daily life by suppressing Sundays. Industrial and administrative workers in each plant were divided up into 5 groups who worked 4 days out of every 5, alternating. The effect of this reorganization was to devastate family and social life. Soviet workers rested more often than their Western counterparts, but they did not rest together and consequently there was no possibility of having an effective public life. The experiment was eventually halted and replaced by a 6-day week: 5 workdays followed by a 6th day of common rest. The Soviet regime gave up the continuous production cycle but did not restore Sundays, so the cultural struggle between modern principles and ancient religious traditions continued. Opposition to the new system then arose from farm workers. They continued to live on the rhythm of the 7-day week: markets were held every 7 days at the time corresponding to the vanished Sunday. Another unfortunate consequence was that the Soviet population split into two distinct societies, which no longer maintained the same time frame: urbanites used the 6-day week, while the rural population rejected it. Finally, the weight of religious tradition prevailed over the reforms: in 1940 the government restored the 7-day week.

Jacqueline de Bourgoing

The Perpetual 'Pataphysical Calendar

The Absurdist playwright Alfred Jarry, inventor of a mysterious form of science or philosophy called 'Pataphysics, has had a wide following since the beginning of the 20th century. His followers wrote a parody of the solemn calendar of the French Revolution.

Founded on May 11, 1948, under the auspices of *The Deeds and Opinions of Doctor Faustroll, Pataphysician,* the College of 'Pataphysics…devoted itself very promptly to "reforming the indescribable misery of traditional calendars." Thus, in Sand 76 His Magnificence the Vice-Curator, Founder of the College, promulgated a new calendar…

This computation takes as its reference September 8, 1873, [Jarry's] birthday. This event marks the beginning of the Pataphysical Era and the beginning of each year (1 Absolu). It is decomposed into 12 months of 28 days, as follows: Absolu [Absolute], Haha, As [Ace], Sable [Sand], Décervelage [Debrainage], Gueules [Mouths], Pédale [Pedal], Clinamen [Allusively Unforeseeable], Palotin [Impaler], Merdre [Crap], Gidouille [Spiral], Tatane [Shoe], and Phalle [Cock]. A 29th imaginary day, called Hunyadi [the name of a Hungarian patriot or a Hungarian laxative], is added to each month and increases to 377 the number of the days of the pataphysical year. This measurement of time gives us the Perpetual 'Pataphysical Calendar, composed and published by the Extraordinary Astrological Rota of the College of 'Pataphysics, which has established the list of the Supreme Holidays. It was finally completed with the addition of *The Lives of Saints,* published in the *Organographs of the Cymbalum Pataphysicum* and then in the *Monitions,* fulfilling the vow of the Transcendant Satrap Boris Vian [a novelist] to endow the College with a Catechism…

Many see in this computation and its glosses an "imaginary solution" to the problem of time, and at the same time the inaugural functional act of a world whose intersections with our own remain frequent.

The year 2000 has only limited significance for the College, which entered its second century in 1977 (vulgar era).

Perpetual 'Pataphysical Calendar, composed and published by the College of 'Pataphysics, Paris, repr. in Jean-Didier Wagneur, *Revue de la Bibliothèque nationale* (*Review of the National Library*), no. 4, January 2000

A modern political calendar

The North African nation of Libya has been governed as the Libyan Arab Republic by Colonel Muammar el-Qaddafi since 1969. A newspaper article describes the difficulties that a nation encounters when it attempts to alter the calendar.

LIBYA UNDER QADDAFI: DISARRAY IS THE NORM

Life can be so unpredictable here that people are not even sure what year it is.

Officially it is 1369. But just two years ago Libya was living through 1429. No one can quite name the day when the count changed, especially since both remain in play.

A newspaper headline last week, for example, heralded the latest legal bulletin, No. 4 for 1431, while the paper itself was dated 1369. Event organizers throw up their hands and put the Western year in parenthesis somewhere on their announcements. Other Libyans just throw up their hands.

"Why do we keep using different dates?" bellowed a woman at her neighborhood Popular Committee meeting, a regular nationwide talkfest about public rules. "Why can't we be like every single other Muslim country and count from the time of the Prophet's migration, not from his death? What is this?"

The unspoken answer is that Libya's leader, Col. Muammar el-Qaddafi, likes it that way. Libyans just want to be normal, but his quixotic decisions keep their lives in an endless state of controlled chaos.

The date is but one example, the leader switching from the standard Muslim calendar to one marking the years since Muhammad's birth and then shifting to one marking his death. The rest of the Islamic world is in 1421, counting from the date the Prophet migrated from Mecca to Medina to found the faith. Colonel Qaddafi also decided some time ago that he disliked both the Western and the Eastern months, so he renamed them. February is Lights. August is Hannibal…

Neil MacFarquhar, *New York Times,* February 14, 2001

Chronological dating systems

The calendar is gradually becoming more international, detached from its original links to religion and local culture. But it is not always as neutral and universal as it seems.

For the sake of convenience in a global society, much of the world has adopted as a universal measure the European system of dividing human history into two eras: BC, or "Before Christ," and AD, *Anno Domini,* following the supposed date of the birth of Jesus. Some historians prefer to use the more neutral terms BCE and CE, "Before the Common Era" and "Common Era," to represent the same two time periods. Other dating systems include the Muslim AH, or *Anno Hegira.* The date AH 0 corresponds to AD 622. The ancient Romans used AUC, *Ab Urbe Condita,* "from the founding of the city of Rome," in 753 BC. Astronomy also relies on the concept of the calendar year as the basis for measuring immense distances. These are calculated according to another universal measure: the distance traveled by light in a vacuum during one year (9,460,528,401,200 km, or 5.8 trillion miles).

But archaeologists, anthropologists, and many other scientists need to date objects and events that are much older than any of our current historically based calendars, whose political and cultural purposes are not appropriate to their work, and whose basic unit of 1 solar year is too short to be useful. Scientific dating techniques such as radiocarbon dating can accurately date an object, or the record of an ancient event such as a volcanic eruption, up to about 40,000 years in the past. The BP,

or Before Present, system records dates in relationship to the present. An international convention established the "present" as AD 1950. Year 0 BP is AD 1950. So a BP date is the number of years before 1950. However, it is not accurate to record a radiocarbon date by subtracting 1950 from the reported age, because a radiocarbon year is not precisely equivalent to a calendar year. For example, a radiocarbon date of 10,000 BP corresponds to a calendar date of about 11,250 to 11,450 years ago.

All these systems are forms of chronometric dating. A chronometric date is one that positions an event chronologically with reference to a universal time scale such as a calendar. Such dates usually are given in terms of the number of years before or after the calendar's starting point, whether it is the birth of Christ or Muhammad's departure from Mecca or some arbitrary moment. Scientists whose work deals with vast time spans have created a different calendar concept: relative dating.

A relative date places the time of an event with reference to other events significant to the discipline being studied or local to the event itself, rather than on a universal scale. It indicates that one event occurred some measurable amount of time earlier or later than another, but it does not provide the means to measure the amount of time in years or centuries or millennia. For instance, a layer of archaeological material deposited on top of another may be determined to be newer than the lower one.

The tyranny of the zeros

We are accustomed to dividing time into centuries. As historians point out, however, time does not always fall easily into these neat packages.

Counting by hundreds

The great triumph of the calendar unit greater than the year is the century, a period of one hundred years. The Romans applied the Latin word *saeculum* to variable periods, often associated with the idea of one generation of humankind. The early Christians, while keeping the ancient meaning of the word, gave it the sense of a human life—that is, earthly life as opposed to the afterlife. But in the 15th century some historians and scholars had the idea of dividing time into segments of one hundred years. The unit is rather long but the figure 100 is simple…The first century in which the word and the term were truly applied was the 18th. From then on this convenient, abstract notion imposed its tyranny on history. Everything thenceforth had to be couched in this artificial mold, as if the century were imbued with life, as if it had a unity, as if things changed from one century to the next.

Jacques Le Goff,
entry for "Calendar" in the
Enciclopedia Einaudi (*Einaudi Encyclopedia*), vol. 2, 1977

Time scales and the year 2000

Writing just before New Year's 2000, the palaeontologist Stephen Jay Gould mused upon our fascination with that date. Now that the date has passed, some of the fears that preceded it have also faded.

What frightens us in our secular age is the computer breakdown that'll occur if computers interpret the 00 of the year 2000 as a return to 1900. But no one dreams of claiming that this threat to computers will be an Apocalypse in the biblical sense…

The year 2000 is a very special date,

it's true, but for reasons that have to do with the history of the calendar. We're going to have a very rare opportunity, you know, not only of witnessing a change of millennium but a change of century with a 29 February in it. Because the year 2000 will be a leap year…The complexity of the calendar is a permanent challenge to human ingenuity…

The year 2000 would occur in the history of the world even if we had a different system of calculation. But we give it psychological significance, for whatever mysterious reason, because the human mind seems to need cycles that have meaning within a mathematical system. In our system, hundreds and thousands only have the meaning we give them. When we talk about the nineteenth century or the twentieth century we are giving a meaning to arbitrary categories.

The year 2000 seems special to us because our system of arithmetic is based on the number ten. Ten is an excellent basis for calculation, with many advantages. There's nothing to prove that its choice is in any way related to the fact that we all have ten fingers, but I would be surprised if there were no connection…However, having ten fingers is no guarantee that you'll end up with a decimal system. The Mayas counted in twenties—they probably counted on their fingers *and* their toes! Theirs was a good mathematical system, with complex cycles analogous to our own, cycles of 1,600 and 2,400 years. In a system like that, of course, the number 2,000 has no particular significance…

Does the new century begin in the year 0 or the year 1? The beginning of our century was celebrated on 1 January 1901, but…the celebrations this time… take place at the turn of the year 2000. So we could say…that our century will

be only ninety-nine years long…

It's an excellent solution to a debate that's been going on for several centuries, and certainly since the turn of the century from 1599 to 1600. It's a trivial and meaningless debate in fact, but everybody gets very excited about it…

When we made time begin again on 1 January 753 AUC ["from the founding of Rome"], that day became 1 January in the year 1. And that's why our centuries begin with 01 and not with 00…The choice of 1 January in the year 1 is simply a convention, it's neither true nor false. But what one's got to realize is that the choice of convention, whatever it may be, has long-term consequences…If you make the calendar begin on 1 January in the year 1, and you're determined that a century absolutely must last 100 years, then the century ends only after the hundredth year. That's the mathematically correct way of looking at things.

On the other hand, given the way we write our numbers, with a mathematical base of ten, the number 100 looks much more interesting than the year 101, and 1900 is more interesting than 1901. Between 1899 and 1900 you change all the numbers except the first. Between 1900 and 1901 you change only one. Between 1999 and 2000 you change all the numbers. Between 2000 and 2001 you change only one. You may say that this is nonsensical, of no importance, that it's trivial, all in the mind. But my answer would be that that's just what a calendar is, a set of small calculations that serve to create a convention that's valid for all of us.

Stephen Jay Gould,
*Conversations about the
End of Time*, 1998,
trans. by Ian Maclean
and Roger Pearson

Johannes de Bry, Calendarium Naturale Magicum Perpetuum (Magical Calendar), Frankfurt, 1620

Further Reading

Achelis, E., *The Calendar for the Modern Age*, 1959.

Aveni, A. F., *Empires of Time: Calendars, Clocks and Cultures*, 1995.

Boorstin, D. J., *The Discoverers: A History of Man's Search to Know His World and Himself*, 1983.

Borst, A., *The Ordering of Time: From the Ancient Computus to the Modern Computer*, trans. A. Winnard, 1993.

Duncan, D. E., *Calendar*, 1998.

Durkheim, E., *The Division of Labor in Society*, 1984.

Eliade, M., *The Myth of the Eternal Return*, trans. W. R. Trask, 1971.

————, *Mystic Stories: The Sacred and the Profane*, trans. A. Cartianu, 1992.

Elias, N., *Time: An Essay*, 1992.

Fraser. J. T., *Time: The Familiar Stranger*, 1987.

Gould, S. J., *Questioning the Millennium : A Rationalist's Guide to a Precisely Arbitrary Countdown*, 1997.

Granet, M., *Chinese Civilization*, 1930.

Irwin, K. G., *The 365 Days*, 1963.

Landes, D. S., *Revolution in Time: Clocks and the Making of the Modern World*, 1983.

Le Goff, J. *Time, Work, and Culture in the Middle Ages*, 1980.

Parise, F., ed., *The Book of Calendars*, 1982.

Zerubavel, E., *The Seven Day Circle: The History and Meaning of the Week*, 1985.

List of Illustrations

Index

Photograph Credits

Text Credits

Jacqueline de Bourgoing holds degrees in geography
and serves on the faculty of the Institute of
Political Studies in Paris. She is active with French
television's History channel, for which she has produced
a series on the history of the calendar.

The author wishes to thank Jacques Le Goff, Claude Gaignebet,
Perrine Mane, and Dominique Missika.

Translated from the French by David J. Baker and Dorie B. Baker

For Harry N. Abrams, Inc.
Editor: Eve Sinaiko
Typographic designers: Elissa Ichiyasu, Tina Thompson
Cover designer: Brankica Kovrlija
Text permissions: Barbara Lyons

Library of Congress Cataloging-in-Publication Data

Bourgoing, Jacqueline.
  [Calendrier. English]
  The calendar : history, lore, and legend / Jacqueline Bourgoing.
    p. cm. — (Discoveries)
  Includes bibliographical references (p.) and index.
  ISBN 0-8109-2981-3 (pbk.)
    1. Calendar—History. 2. Calendar—Folklore. 3. Calendar—History—
  Pictorial works. 4. Time—Social aspects—History.   I. Title. II. Discoveries
  (New York, N.Y.)
CE6 .B6813 2001
529'.3'09—dc21   2001022409

Printed and bound in Italy by Editoriale Lloyd, Trieste